Competence in High Performance Computing 2010

Christian Bischof • Heinz-Gerd Hegering
Wolfgang E. Nagel • Gabriel Wittum
Editors

Competence in High Performance Computing 2010

Proceedings of an International Conference on Competence in High Performance Computing, June 2010, Schloss Schwetzingen, Germany

Springer

Editors
Christian Bischof
Technical University Darmstadt
Mornewegstr. 30
64293 Darmstadt
Germany
christian.bischof@tu-darmstadt.de

Heinz-Gerd Hegering
Leibniz Rechenzentrum (LRZ)
Boltzmannstr. 1
85748 Garching
Germany
hegering@lrz.de

Wolfgang E. Nagel
Technical University Dresden
Center for Information Services
and High Performance Computing (ZIH)
01062 Dresden
Germany
wolfgang.nagel@tu-dresden.de

Gabriel Wittum
Goethe University Frankfurt
Goethe Center for Scientific Computing
Kettenhofweg 139
60325 Frankfurt
Germany
wittum@gcsc.uni-frankfurt.de

Front cover figure: Groundwater flow around a tunnel system in porous fractured rock at the Äspö Hardrock Laboratory. Left: coarse grid on 64 processors, right: pressure distribution with 10 selected iso-surfaces. Geometry and hydrogeologic data courtesy of Swedish Nuclear Fuel and Waste Management Co (SKB), Stockholm, Sweden and Gesellschaft für Anlagen- und Reaktorsicherheit (GRS) mbH, Braunschweig, Germany. By Courtesy of Dr. A. Nägel, G-CSC

ISBN 978-3-642-44035-9   ISBN 978-3-642-24025-6 (eBook)
DOI 10.1007/978-3-642-24025-6
Springer Heidelberg Dordrecht London New York

Mathematics Subject Classification (2010): 65Cxx, 65C99, 68U20

Printed on acid-free paper

Springer is part of Springer Science+Business Media (www.springer.com)

Foreword

This volume contains a blend of papers focusing on high performance computing. The papers were contributed to the conference "Competence in High performance Computing" which took place in Schwetzingen Castle from June 22 to 24, 2010. The conference was the first one in a series of annual conferences of the newly founded Gauß-Allianz of the High Performance Computing Centers in Germany. One of the major topics was the presentation of projects funded by BMBF in the HPC program. The range of the papers reaches from an overview over progress in methods for HPC like parallel multigrid methods for large systems of equations to questions like performance measurements and energy efficient cluster computing.

Special thanks are due to Eva-Maria Vachek, who carefully compiled the contributions to this volume.

Darmstadt	*Christian Bischof*
Munich	*Heinz-Gerd Hegering*
Dresden	*Wolfgang E. Nagel*
Frankfurt	*Gabriel Wittum*

Contents

HICFD: Highly Efficient Implementation of CFD Codes for HPC Many-Core Architectures

Achim Basermann, Hans-Peter Kersken, Andreas Schreiber, Thomas Gerhold, Jens Jägersküpper, Norbert Kroll, Jan Backhaus, Edmund Kügeler, Thomas Alrutz, Christian Simmendinger, Kim Feldhoff, Olaf Krzikalla, Ralph Müller-Pfefferkorn, Mathias Puetz, Petra Aumann, Olaf Knobloch, Jörg Hunger, and Carsten Zscherp

Abstract The objective of the German BMBF research project *Highly Efficient Implementation of CFD Codes for HPC Many-Core Architectures* (HICFD) is to develop new methods and tools for the analysis and optimization of the performance

---

A. Basermann (✉) · A. Schreiber
German Aerospace Center e.V. (DLR), Simulation and Software Technology, Cologne, Germany
e-mail: achim.basermann@dlr.de; hans-peter.kersken@dlr.de; andreas.schreiber@dlr.de

T. Gerhold · J. Jägersküpper · N. Kroll
DLR, Institute of Aerodynamics and Flow Technology, Göttingen, Germany
e-mail: thomas.gerhold@dlr.de; jens.jaegerskuepper@dlr.de; norbert.kroll@dlr.de

J. Backhaus · H.-P. Kersken · E. Kügeler
DLR, Institute of Propulsion Technology, Cologne, Germany
e-mail: jan.backhaus@dlr.de; edmund.kuegeler@dlr.de

T. Alrutz · C. Simmendinger
T-Systems Solutions for Research GmbH, Bunsenstr. 10, D-37073 Göttingen, Germany
e-mail: thomas.alrutz@t-systems-sfr.com; christian.simmendinger@t-systems-sfr.com

K. Feldhoff · O. Krzikalla · R. Müller-Pfefferkorn
Technische Universität Dresden, Zentrum für Informationsdienste und Hochleistungsrechnen, Zellescher Weg 12, D-01062 Dresden, Germany
e-mail: kim.feldhoff@tu-dresden.de; olaf.krzikalla@tu-dresden.de; ralph.mueller-pfefferkorn@tu-dresden.de

M. Puetz
IBM Deutschland GmbH, Hechtsheimer Str. 2, D-55131 Mainz, Germany
e-mail: mathias.puetz@de.ibm.com

P. Aumann · O. Knobloch
Airbus Deutschland GmbH, Aerodynamic Tools and Simulation, Bremen, Germany
e-mail: petra.aumann@airbus.com; olaf.knobloch@airbus.com

J. Hunger · C. Zscherp
MTU Aero Engines GmbH, Dachauer Str. 665, D-80995 München, Germany
e-mail: joerg.hunger@mtu.de; carsten.zscherp@mtu.de

C. Bischof (eds.), *Competence in High Performance Computing 2010*,
DOI 10.1007/978-3-642-24025-6_1,

of parallel computational fluid dynamics (CFD) codes on high performance computer systems with many-core processors. In the work packages of the project it is investigated how the performance of parallel CFD codes written in C can be increased by the optimal use of all parallelism levels. On the highest level Message Passing Interface (MPI) is utilized. Furthermore, on the level of the many-core architecture, highly scaling, hybrid OpenMP/MPI methods are implemented. On the level of the processor cores the parallel Single Instruction Multiple Data (SIMD) units provided by modern CPUs are exploited.

1 Introduction

The research project HICFD is funded by the German Ministry for Education and Research within the programme *IKT 2020 – Research and Innovation*. The project's objective is to develop new methods and tools for the analysis and optimization of the performance of parallel computational fluid dynamic (CFD) codes on high performance computer systems with many-core processors and to exemplarily apply these to DLR's CFD programs TAU [11] for the computation of external flows and TRACE (Turbo-machinery Research Aerodynamic Computational Environment) [12] for the simulation of internal flows.

In the workpackages of the project it is examined how the performance of parallel CFD codes can be increased by the optimal exploitation of all parallelism levels. On the highest, with Message Passing Interface (MPI) parallelized level an intelligent mesh partitioning in order to improve the load balance between the MPI processes is promising. For the block-structured grids used in TRACE a many-core compatible partitioning tool is developed.

Furthermore, on the level of the many-core architecture, highly scaling, hybrid OpenMP/MPI methods (OpenMP: Open Multi-Processing) are implemented for the CFD solvers TAU and TRACE. Within the block-structured CFD code TRACE the iterative algorithm for the solution of linear systems of equations has to be optimized, among other things by preconditioning methods adequate for many-core architectures.

On the level of the processor cores a pre-processor is developed which makes a comfortable exploitation of the parallel Single Instruction Multiple Data (SIMD) units possible, also for complex applications. For a detailed performance examination of SIMD operations the tracing abilities of the performance analysis suite VAMPIR *Visualization and Analysis of MPI Resources* [13] is further developed.

2 A Generic SIMD Preprocessor

Data parallelism at the level of the processor cores using their SIMD units cannot be exploited simply by hand. First, using SIMD instructions makes the code highly platform dependent. Second, data-level parallel programming using SIMD instructions is not a simple task. SIMD instructions are assembly-like low-level and often steps like finalization computations necessary for the vectorized execution of loops. Thus it is essential to find or develop a tool suitable to comfortably and automatically vectorize CFD codes.

A vectorization tool is best built in a compiler. All current C compilers provide autovectorization units. We tested the compilers being in consideration for the HICFD project wrt. their autovectorization capabilities but had to decide not to rely on them regarding the vectorization process due to the following reasons:

- Each compiler has its own means to provide meta information. It is impossible to hide the different syntactic forms in a catch-all macro since some annotations like pragmas cannot be generated by the C preprocessor. Thus, in order to gain the required relative independence from a particular compiler we would have to pollute the source code with a lot of vendor-dependent annotations.
- During our tests a lot of subtle issues arose around compiler-generated vectorization. For instance in one case a compiler suddenly rejected the vectorization of a particular loop just when we changed the type of the loop index variable from `unsigned int` to `signed int`. A compiler expert can often reason about such subtleties and can even dig in the documentation for a solution. An application programmer, however, normally concentrates on the algorithms and cannot put too much effort in the peculiarities of each compiler used.
- The vectorization of certain (often more complex) loops was rejected by all compilers regardless of inserted pragmas, given command-line options etc.

We checked other already available vectorization tools [4,7], but the development focus of these tools did not match our requirements. Thus eventually we decided to develop a new tool in order to comfortably exploit the parallel SIMD units. Based on the requirements imposed by the numerical kernels of the CFD codes and based on experiences collected in former projects [6] the main function of the tool is the vectorization of loops. Beyond the specific target of CFD codes, the tool is usable as a universal vectorizer and aims at becoming an industrial-strength vectorizing preprocessor.

2.1 The Vectorizing Source-to-Source Transformator Scout

We opted for a strict semi-automatic vectorization. That is, as with compilers, the programmer has to mark the loops to be vectorized manually. Thus the analysis phase can be limited to the minimum necessary for vectorization and the focus can be put on code generation. We have called this vectorization tool Scout.

Most properties of Scout are direct consequences of the demands of the HICFD project:

- The input language is C since this is the programming language of the CFD codes considered in HICFD.

- The output language is C, too. This comes due to the fact that the simulation based on the CFD codes has to run on heterogeneous platforms each with different C compilers available.
- Scout focuses on short SIMD architectures. These architectures are very common in the current processor cores of even high performance computer systems.
- Scout is configurable in order to support the various already existing SIMD instruction sets as well as upcoming ones.
- Scout is usable on a day-to-day basis in the software production process. That is, at first a programmer can experiment with the tool in order to find the best transformation possible. In addition he is able to check the correctness of the output with a graphical user interface. Secondly, Scout exposes a command line interface as well and thus can be used as part of an automatic build chain.

2.2 Practical Results

We have applied Scout to two different CFD production codes used in the German Aerospace Center (DLR). Both codes are written using the usual array-of-structure approach. That approach is rather unfriendly wrt. vectorization because vector load and store operations need to be composite. Nevertheless we did not change the data layout but only augmented the source code with the necessary Scout pragmas. Thus the presented performance results were generated using the source code as is. All presented measurements were done on an Intel® Core™ 2 Duo P8600 processor with a clock rate of 2.4 MHz, operating under Windows 7™ using the Intel® compiler version 11.1. Other systems showed similar results, however the GNU compiler 4.4.1 appeared to optimize the SSE intrinsics differently and thus produced slightly slower code.

The first code, TRACE, uses a structured grid. Direct indexing is used in loops, and array indices are linearly transformed loop indices. We extracted four different computation kernels from the code and measured each individually. Kernel 1 computes the viscous fluxes of the RANS equations over a cell area of the control volume while kernel 2 determines the corresponding convective fluxes. In kernel 3, the derivatives of the fluxes according to the state vector (density, velocity etc.) are calculated due to the implicit time discretization of the RANS equations. Kernel 4 solves the equation system resulting from the implicit discretization with the Backward-Euler scheme by an incomplete LU decomposition of the matrix and a subsequent Gauss-Seidel iteration. The four kernels contribute to the wall clock time of a TRACE run with distinctly more than 50%. Figure 1 shows typical speedup factors of the vectorized kernels produced by Scout compared to the original kernels. On the right, the speedup using double precision is shown (two vector lanes), on the left, the speedup using single precision is shown (four vector lanes). The overall picture does not change for larger problem sizes. Usually the number of iterations are roughly problem size cubed since most computations are done in triply nested loops.

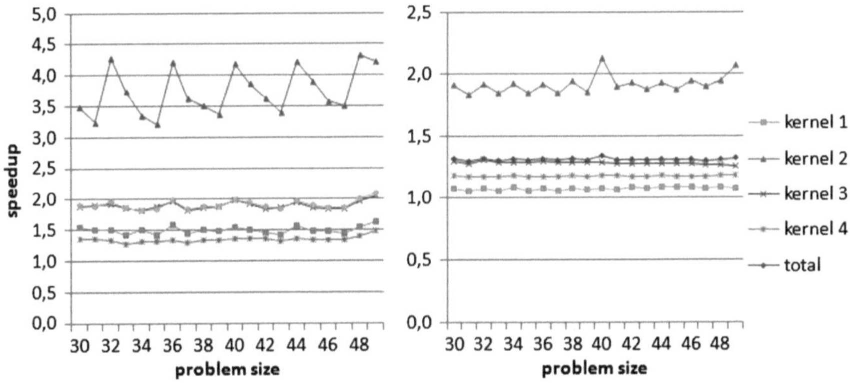

Fig. 1 Speedup of TRACE kernels gained due to the vectorization performed by Scout

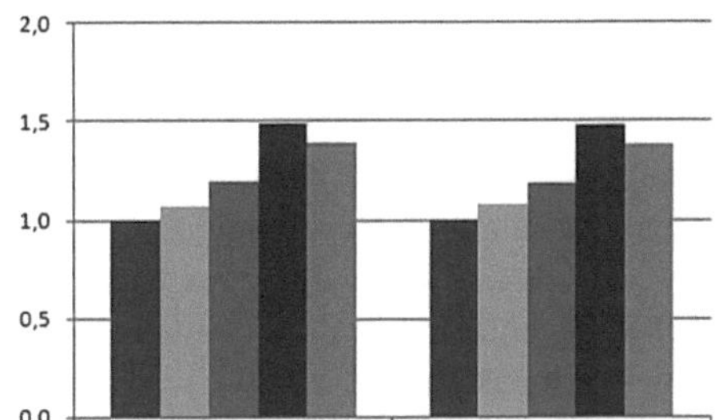

	Grid 1	Grid 2
original	1,000	1,000
merged	1,070	1,075
vectorized	1,194	1,189
merged, vectorized	1,489	1,484
merged, vectorized, compared to merged	1,391	1,381

Fig. 2 Speedup of a partially vectorized TAU kernel

As expected, we gain a higher speedup with more vector lanes since more computations can be executed in parallel. The speedup of each particular kernel heavily depends on the computation workload compared to the number of load and store operations. Kernel 2 even outperforms its theoretical maximum speedup which is a result of further transformations (in particular function inlining) performed by Scout implicitly.

Unlike TRACE, TAU exploits unstructured grids. Thus the loops mostly use indirect indexing to access array data elements. We concentrated our efforts on the most performance critical kernel. The loops in that kernel could only partially be vectorized and to our best knowledge there is currently no other tool capable of vectorizing these loops automatically. Nevertheless we could achieve some speedup as shown in Fig. 2. We had two different grids as input data to our disposal. First we vectorized the code as is. However the gained speedup of about 1.2 was not satisfying. Hence, we merged some loops inside the kernel together to remove repeated traversal over the indirect data structures. This not only resulted in a better performance of the scalar code but also in a much better acceleration due to vectorization.

3 Performance Analysis of SIMD Codes

In the following, the performance of the parallel CFD programs TAU and TRACE with respect to the application of SIMD techniques for code acceleration are examined. By means of SIMD techniques, a single machine instruction can, at the same time, be applied to multiple data elements.

In order to investigate where and when a (parallel) program makes use of SIMD methods a special benchmark suite was developed and the performance analysis software VAMPIR was extended to visualize and analyze the SIMD operations. With the help of graphical VAMPIR displays, a user is able to verify where the SIMD preprocessor Scout (cf. Sect. 2) produces source code on the basis of SIMD techniques. In detail, the development of the displays targets answering the following questions: Are SIMD techniques used? If so where are they used? At which scale SIMD techniques are used? How can the quality of the application of SIMD techniques be measured?

Detailed information about the application of SIMD techniques during a program run can be collected by exploiting the information stored in hardware performance counters. These counters are special registers on microprocessors, in which values about hardware events (the total number of all executed SIMD instructions, e.g.) are stored.

For the development of the new graphical VAMPIR displays, it is advantageous to first analyze the SIMD performance of a program on a coarse-grained level. Therefore, a benchmark suite designed to statistically measure the SIMD performance of a program based on the information stored in hardware performance counters was developed. This SIMD benchmark suite (called *simdperfana*) consists of a C library (called *simdbench*) and a (*bash*) shell script collection (called *ccperfsimd*). SIMD instructions are recorded by means of hardware performance counters which will be accessed in the high level programming language C via PAPI (*Performance API*) [10].

simdbench and *ccperfsimd* interact in the following way: By means of the C library *simdbench*, a user can instrument source code parts of a program. By means of the shell script collection *ccperfsimd*, the SIMD performance of the instrumented program is automatically measured (program runs with different compiler options: vectorization type, optimization level, SIMD technique).

The benchmark suite is slim, clearly arranged and easy to use: *simdbench* offers four C macros (prolog, start, stop, epilog) to trigger a measurement, measurement parameters are changeable via *bash* environment variables. Furthermore, *simdperfana* supports PAPI and VampirTrace (a tool for tracing program runs which works together with VAMPIR) [14]. Thus the SIMD performance can easily be analyzed with VAMPIR.

On the basis of various tests (e.g. program runs of different numerical kernels), a VAMPIR display was adapted so that a user can detect where a program makes use of SIMD techniques and at which scale they are used.

4 Hybrid Shared-Memory/MPI Parallelization

In the modern process of the aerodynamic design of aircrafts or turbines for jet engines, CFD plays a key role. Large-scale computing resources allow more detailed numerical investigations with bigger numerical configurations. Up to 50 million grid points in a single simulation are becoming a standard configuration in the industrial design. These requirements demand a shift from the paradigm of *single-program single-data (SPSD)* per MPI-Rank to a more sophisticated approach with *single-program multiple-data (SPMD)* per MPI-Rank. The benefit and techniques of this hybrid parallelization approach are demonstrated with the two DLR CFD solvers TAU and TRACE.

4.1 Hybrid Multistage Shared-Memory/MPI Parallelization of the TAU Flow Solver

At project start the parallelization of the TAU solver was based on MPI using a vertex-oriented domain decomposition per process with one layer overlap between the domains for the MPI communication. Typical simulations with TAU for three dimensional geometries use grid sizes between 10 and 50 million grid points and are computed with 32–512 MPI processes. Usually, such computations are performed employing a geometrical multigrid method with a 3- or 4-level W multigrid cycle. For e.g. a grid with 30 million grid points and 512 MPI processes, each grid domain contains approximately 60,000 grid points (without overlap) on the finest grid and 7,500, 940, and 120 points on the second, third, and forth grid level, respectively. For such applications TAU has a very good parallel efficiency as can be seen in Fig. 3.

This figure also shows that, at least for a 4-level W multigrid cycle, the parallel efficiency decreases significantly when using more than 2,000 MPI processes (see also [2]). The degradation of the efficiency is caused by the effect that the grid partitions become very small, especially on the coarse multigrid level so that the number of grid points in the overlap layer is no longer negligible. For instance with 4,096 partitions, the number of additional points to be communicated is already about 35% of the total number of points in the non-decomposed grid. This effect is even worse on the coarse multigrid levels. One possibility to avoid such small grid partitions (for this number of processes) is a hybrid shared-memory/MPI

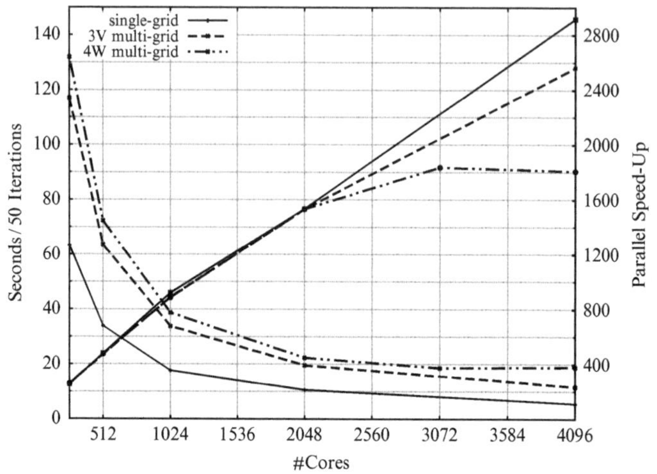

Fig. 3 TAU-Code performance up to 4,096 cores on AMD Barcelona 1.9 GHz with Infiniband DDR ConnecX interconnect. Strong scaling for a mesh with 31 million points, cell-vertex metric, one MPI process per core. Explicit Runge/Kutta RANS solver with multi-grid (pre-smoothing only; 3V/4W: 3-level V/4-level W cycle): central discretization, scalar dissipation, Spalart/Allmaras turbulence model

parallelization. Following this concept, the grid is partitioned into as many domains as cluster nodes (or sockets) are to be used (rather than the number of cores). This increases the grid sizes per partition by one order of magnitude as today's clusters have nodes with 8 to 24 cores. Thus, the potential of this strategy is to improve the (strong) scalability, i.e., to shift the decrease of parallel efficiency to much higher core numbers.

Therefore, a prototype version of TAU has been implemented using multiple threads per cluster node (one per core) for the flow computation inside the grid domain. Between the grid domains, data is exchanged using MPI as before. The main computational work in the solver is in loops over points and edges of the grid. Points and edges of each grid domain are grouped into sub-domains, called colors, which are used to optimize the cache access. The computation of these colors is performed in parallel by multiple threads. As there are points that belong to more than one color, the updates of these points have to be serialized to avoid data races (when different threads update the same point at the same time). For this serialization, a locking mechanism for the colors has been developed to ensure mutual exclusion of neighboring colors. Namely, at any point in time, all colors processed (in parallel by multiple threads) do not share any points. The computation of the colors is asynchronous except at some synchronization points when MPI communication with neighboring domains has to be done. Performance benchmarks for the hybrid-parallel implementation are ongoing work; preliminary results are encouraging (cf. [2]).

4.2 Hybrid Shared-Memory/MPI Parallelization of the TRACE Flow Solver

In the design phase of multistage compressors and turbines for jet engines, or stationary gas turbines, complete computations of the configurations over night are mandatory for the engineers in order to analyze and improve the design throughout the next business day.

At the Institute for Propulsion Technology of the German Aerospace Center (DLR), the parallel simulation system TRACE has been developed specifically for the calculation of internal turbo-machinery flows. TRACE applies a finite volume approach with block-structured grids for solving the associated Navier-Stokes equations. In the MPI parallelization of the code one MPI process is assigned to each block. Thus the scalability of TRACE is limited to the number of blocks. Moreover the load balance is dominated by the largest block. While it is possible to split the largest block into smaller ones, it is often not advisable to do so since smaller MPI domains have a larger overhead (due to the overlap region) and also decrease the convergence rate of an implicit solver. Hence a hybrid parallelization is the matter of choice in order to improve the performance and scalability of the TRACE code.

The implementation of the hybrid parallelization concept made a complete redesign of the TRACE data structures necessary. In order to maximize spatial and temporal data locality the data structures of TRACE were reorganized into hot and cold parts. The employment of a novel hyperplane formulation of the Gauss-Seidel relaxation algorithm, where the respective elements were addressed directly in hyperplane coordinates, makes it possible to improve also the scalar efficiency by a considerable margin. Speedup factors of up to 2.4 were observed, depending on the use-case and hardware platform.

The hybrid parallelization concept itself is based on pthreads and MPI. In this implementation pthreads are used for all processing units of a CPU socket and the enveloping MPI process is bound to the socket.

With these improvements an engineer is now able to compute a complete compressor map over night. The example given in Table 1 is a 15 stage compressor with an additional inlet guide vane and an outlet guide vane. The mesh has 19,11 million grid points overall for 32 blade rows. Table 1 shows the runtime for a single point on the compressor map. If Hyperthreading is enabled a speedup factor of 2.7 on the 4 cores of an Intel Xeon 5540 (Nehalem) is obtained by the shared-memory parallelization introduced (see also [9]). The hybrid parallelization made the efficient use of a much higher node number for the compressor problem possible so that the execution time could be reduced from 6 h 45 min (pure MPI) to 2 h 31 min (hybrid concept), cf. Table 1 for details.

Table 1 Runtime comparison for the compressor use-case

Testcase	Compiler	MPI	# nodes	# MPI processes	# threads /MPI process	Runtime in h
Compressor	icc 11.1	Openmpi	8	60	1	06:45:30
Compressor	icc 11.1	Openmpi	30	60	8	02:31:46

5 Scalable Algorithms for Many-Core Architectures

In order to increase the scalability of CFD solvers with multi-block grids like TRACE the development of a generic partitioning tool for block-structured grids is required (cf. Sect. 5.1). On the other hand it is necessary to improve the parallel efficiency of the linearized TRACE solvers by scalable preconditioning for the iterative solution method for sparse systems of linear equations (cf. Sect. 5.2).

5.1 Partitioning for Block-Structured Grids

A promising approach to increase the scalability of parallel CFD solvers with block-structured grids on the highest, with MPI parallelized level is an intelligent grid partitioning for improving the load balance between the MPI processes. For TRACE, we developed a many-core compatible partitioning tool which will be available as open-source library after the end of the project.

A first version was already integrated into TRACE and exploits ParMETIS [5] for partitioning the block graph. The nodes of this graph are weighted with the computational load per block while the edge weights represent the data exchange volume between the block faces. The advantage compared with the standard block distribution in TRACE which just considers the calculations per block is a reduction of the communication overhead. Figure 4, left, shows that this advantage increases with increasing core number for TRACE sample runs on DLR's AeroGrid cluster (45 Dual-processor nodes; Quad-Core Intel Harpertown; 2.83 GHz; 16 GB main memory per node; InfiniBand interconnection network between the nodes).

Another main source for scalability losses in TRACE is the significant size difference of the blocks. Therefore, the partitioning tool developed will be extended by a splitting feature of large blocks into smaller ones. This makes on the one hand a more advantageous load balance and on the other hand the efficient exploitation of a higher core number possible. Figure 4, right, illustrates this for the case that the largest of 255 blocks is manually split into halves for TRACE sample runs on the AeroGrid cluster. The resulting 256 blocks are then distributed with the new partitioner.

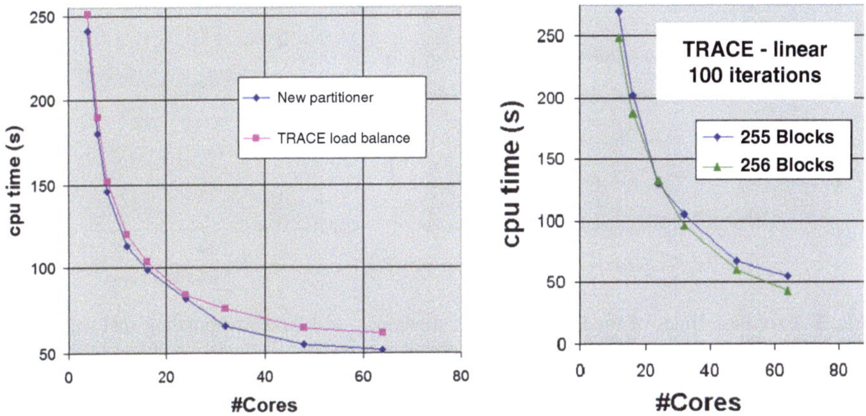

Fig. 4 Execution times of TRACE sample runs with the new partitioner and the standard TRACE load balancer (*left*) as well as with and without manual splitting of the largest block using the new partitioner (*right*)

5.2 *Preconditioning for the Linear TRACE Solvers*

For the parallel iterative solution of sparse equation systems to be solved within the linearized TRACE solvers, FGMRes with Distributed Schur Complement (DSC) preconditioning [8] for real or complex matrix problems has been investigated.

The DSC method requires adequate partitioning of the matrix problem since the order of the approximate Schur complement system to be solved depends on the number of couplings between the sub-domains. Graph partitioning with ParMETIS [5] from the University of Minnesota is suitable since a minimization of the number of edges cut in the adjacency graph of the matrix corresponds to a minimization of the number of the coupling variables between the subdomains. The latter determine the order of the approximate Schur complement system used for preconditioning.

For the solution of TRACE linear equation systems, we developed a parallel iterative FGMRes algorithm with DSC preconditioning. In [3] and [1], we demonstrated the superiority of the DSC preconditioner over (approximate) block-Jacobi preconditioning. Block-local preconditioning methods like block-Jacobi are the standard preconditioners in TRACE. Here, we particularly discuss numerical and performance results of DSC methods for typical complex TRACE CFD problems on many-core architectures.

Figure 5, left, displays execution times of the real and complex DSC solver software developed on 4–32 processor cores of the AeroGrid cluster for a complex and the corresponding real small TRACE matrix problem of order 28,120 or 56,240. The FGMRes iteration was stopped when the current residual norm divided by the initial residual norm was smaller than 10^{-5}. The complex DSC solver version distinctly outperforms the real version. This is caused by the lower problem order

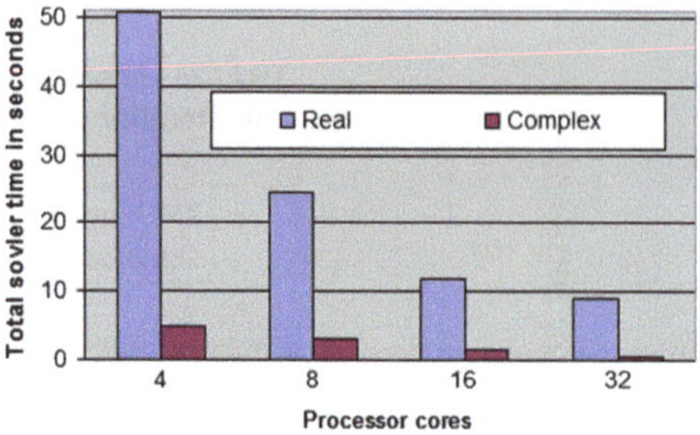

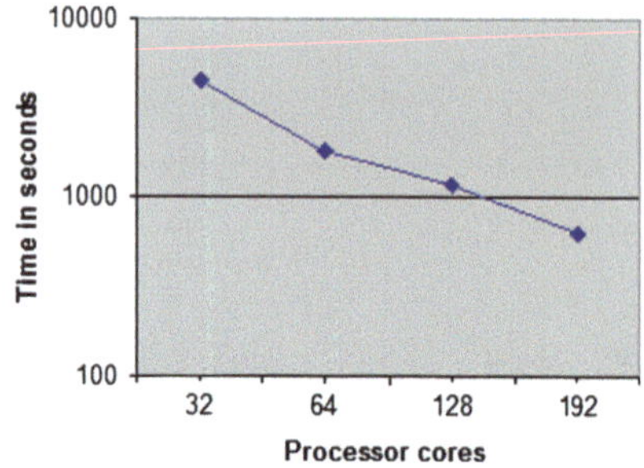

Fig. 5 Execution times of the DSC solver for a complex and the corresponding real small size TRACE matrix problem (*left*) as well as for a complex large TRACE matrix problem (*right*) on a many-core cluster

and a more advantageous matrix structure in the complex case opposite to the real case. In addition, the complex formulation results in higher data locality (storage of real and imaginary part in adjacent memory cells) and a better ratio of computation to memory access due to complex arithmetics in comparison with the real formulation. Figure 5, left, also shows an advantageous strong scaling behavior of the DSC method on the many-core cluster, even for this small matrix problem.

Figure 5, right, shows execution times of the complex DSC method on 4–192 processor cores of the AeroGrid cluster for a large complex TRACE matrix problem of order 4,497,520 and with 552,324,700 non-zeros. In order to achieve the required accuracy in each component of the solution vector the FGMRes iteration was here stopped when the current residual norm divided by the initial residual was smaller than 10^{-10}. Figure 5 demonstrates that the complex DSC algorithm scales very well on many-core clusters for different CFD matrix problems of varying size.

6 Conclusions

After two of three HICFD project years, the methods and tools developed showed promising performance results for the targeted CFD codes TAU and TRACE on all three parallelism levels considered. Significant speedups could be gained by the optimized exploitation of SIMD operations on the processor core level, hybrid parallelization techniques on the level of the many-core architecture and more scalable algorithms on the highest, with MPI parallelized level.

In the remaining project time, the developments on all parallelism levels will be completed and fully integrated into TAU and TRACE. An evaluation with industrially relevant test cases will demonstrate the increased competitiveness of the CFD simulations codes modified in HICFD.

Acknowledgements This work has been supported by the German Federal Ministry of Education and Research (BMBF) under grant 01IH08012 A.

References

1. Alrutz, T., Aumann, P., Basermann, A. et al.: HICFD – Hocheffiziente Implementierung von CFD-Codes für HPC-Many-Core-Architekturen. In: Mitteilungen – Gesellschaft für Informatik e. V., http://www.fg-pars.gi-ev.de/fileadmin/gliederungen/fb-ti/fg-pars/Workshops/PARS/2009/PARS-Mitteilungen_2009.pdf, Parallel-Algorithmen und Rechnerstrukturen, ISSN 0177 - 0454, pp. 27–35 (2009)
2. Alrutz, T., Simmendinger, C., Gerhold, T: Efficiency enhancement of an unstructured CFD-Code on distributed computing systems. In: Parallel Computational Fluid Dynamics, Recent Advances and Future Directions, DEStech Publications, Inc., Lancaster, PA, USA (2010)
3. Basermann, A., Cortial-Goutaudier, F., Jaekel, U., Hachiya, K.: Parallel solution techniques for sparse linear systems in circuit simulation. In: Proceedings of the 4th International Workshop on Scientific Computing in Electrical Engineering, Series: Mathematics in Industry, ISBN 3-540-21372-4, Springer Berlin Heidelberg, Germany (2004)
4. Hohenauer, M., Engel, F., Leupers, R., Ascheid, G., Meyr, H.: A SIMD optimization framework for retargetable compilers. ACM Trans. Archit. Code Optim. **6**(1), Article No. 2 (27 pages), (2009) doi: 10.1145/1509864.1509866
5. Karypis, G., Kumar, V.: ParMETIS: Parallel graph partitioning and sparse matrix ordering library. Tech. rep. # 97-060, University of Minnesota (1997)
6. Müller-Pfefferkorn, R., Nagel, W.E., Trenkler, B.: Optimizing Cache Access: A Tool for Source-to-Source Transformations and Real-Life Compiler Tests. In: Danelutto, M., Vanneschi, M., Laforenza, D. (eds.) Lecture Notes in Computer Science **3149**, pp. 72–81. Springer, Heidelberg (2004)
7. Pokam, G., Bihan, S., Simonnet, J., Bodin, F.: SWARP: a retargetable preprocessor for multimedia instructions. Concurr. Comput.: Pract. Exper. **16**(2–3), pp. 303–318, (2004) doi: 10.1002/cpe.v16:2/3
8. Saad,Y., Sosonkina, M.: Distributed Schur complement techniques for general sparse linear systems. SISC **21**, 1337–1356 (1999)
9. Simmendinger, C., Kügeler, E.: Hybrid Parallelization of a Turbomachinery CFD Code: Performance Enhancements on Multicore Architectures. In: Proceedings of the V European Conference on Computational Fluid Dynamics ECCOMAS CFD 2010, J.C.F. Pereira, A. Sequeira and J.M.C. Pereira (Eds), Lisbon, Portugal, 14–17 June 2010, CD-ROM, ISBN: 978-989-96778-1-4
10. Performance Application Programming Interface: http://icl.cs.utk.edu/papi/.Cited15Dec2010
11. DLR Institute of Aerodynamics and Flow Technology: http://www.dlr.de/as/.Cited15Dec2010
12. DLR Institute of Propulsion Technology, Numerical Methods: http://www.dlr.de/at/desktopdefault.aspx/tabid-1519/2123_read-3615/.Cited15Dec2010
13. Vampir - Performance Optimization: http://www.vampir.eu/.Cited15Dec2010
14. VampirTrace: http://tu-dresden.de/die_tu_dresden/zentrale_einrichtungen/zih/forschung/software_werkzeuge_zur_unterstuetzung_von_programmierung_und_optimierung/vampirtrace/.Cited5Oct2011

Effective Holistic Performance Measurement at Petascale Using IPM

Karl Fürlinger, Nicholas J. Wright, David Skinner, Christof Klausecker, and Dieter Kranzlmüller

Abstract As supercomputers are being built from an ever increasing number of processing elements, the effort required to achieve a substantial fraction of the system peak performance is continuously growing. Tools are needed that give developers and computing center staff holistic indicators about the resource consumption of applications and potential performance pitfalls at scale. To use the full potential of a supercomputer today, applications must incorporate multilevel parallelism (threading and message passing) and carefully orchestrate file I/O. As a consequence, performance tools must also be able to monitor these system components in an integrated way and at the full machine scales. We present IPM, a modularized monitoring approach for MPI, OpenMP, file I/O, and other event sources.

1 Introduction

Supercomputing systems are being built with an ever increasing number of processing elements. Doubling roughly every thirteen months, the performance increase of supercomputers has outpaced the predictions of Moore's law, primarily due to

K. Fürlinger (✉)
University of California at Berkeley, Berkeley, CA, USA
e-mail: fuerling@eecs.berkeley.edu

N.J. Wright · David Skinner
NERSC, Lawrence Berkeley National Laboratory, 1 Cyclotron Road, Berkeley, CA 94720, USA
e-mail: njwright@lbl.gov; deskinner@lbl.gov

C. Klausecker
Ludwig-Maximilians-Universität München (LMU), Munich, Germany
e-mail: klauseck@nm.ifi.lmu.de

D. Kranzlmüller
Ludwig-Maximilians-Universität München (LMU) and Leibniz Supercomputing Centre (LRZ), Munich, Germany
e-mail: kranzlmueller@ifi.lmu.de

C. Bischof (eds.), *Competence in High Performance Computing 2010*,
DOI 10.1007/978-3-642-24025-6_2,

increased level of parallelism [19]. This trend will only be exacerbated by the widespread adoption of multi- and manycore CPUs in the future. In fact, exascale systems are predicted to feature millions of compute cores. It is therefore clear that some form of multilevel parallelism will play a major role in the programming of these systems. A second trend is decreased overall I/O capability. The ratio of I/O bandwidth to FLOPS is decreasing and in an exascale timeframe are expected to be at least an order of magnitude worse, relatively speaking, than today. Therefore I/O performance is becoming an increasingly important factor. To understand all these potential complex performance issues and their interactions there is a clear need for tools that will allow application developers to gain an understanding of their performance issues.

In previous work we have described our solution to this problem – the Integrated Performance Monitoring (IPM) framework [15, 20]. To date, primarily because of our focus on ease of use, more than 310K batch job performance profiles have been collected on NERSC machines over the past 6 years. This focus on ease of use and providing a compact overall view of performance has also led to IPM being used in several commercial HPC vendors to allow them to gain an understanding of their potential customers codes within the context of procurements. The collected IPM performance profiles provide a rich data set for exploring topics of interest of managers of supercomputing centers. By analyzing the data they contain about memory, compute and network usage it is possible to make vital decisions about where to provision more resources to remove bottlenecks, as well as allowing the determination of attractive features for future machine procurements. For example, with the current pressures on memory per core, an understanding of an HPC center's workloads memory requirement is crucial.

In this paper we present a re-architected implementation of our existing workload and performance analysis tool IPM which features a modularized design and adds monitoring modules for OpenMP, and file I/O operations. IPM retains the focus on ease of use and avoiding application perturbation while focusing less on a drill-down into the applications than other tools.

The rest of this paper is organized as follows: In Sect. 2 we introduce the design and implementation of IPM. We describe the event data sources, the event processing and storage, and the data processing and analysis. In Sect. 3 we evaluate the efficiency of IPM. We describe related work in Sect. 4 and conclude in Sect. 5 with an outlook on future work.

2 Performance Monitoring with IPM

2.1 IPM Design Principles

The objective of IPM is to deliver an inventory of program execution events in sufficient detail to inform the user about performance while introducing minimal application overheads. The general model we assume is that of an application

<table>
<tr><td colspan="10">0</td><td colspan="10">0</td><td colspan="10">0</td><td colspan="2">0</td></tr>
<tr><td>0</td><td>1</td><td>2</td><td>3</td><td>4</td><td>5</td><td>6</td><td>7</td><td>8</td><td>9</td><td>0</td><td>1</td><td>2</td><td>3</td><td>4</td><td>5</td><td>6</td><td>7</td><td>8</td><td>9</td><td>0</td><td>1</td><td>2</td><td>3</td><td>4</td><td>5</td><td>6</td><td>7</td><td>8</td><td>9</td><td>0</td><td>1</td></tr>
<tr><td colspan="12">Event ID</td><td colspan="12">Region ID</td><td colspan="8">Thread ID</td></tr>
<tr><td colspan="16">Callsite ID</td><td colspan="8">Res. Select</td><td colspan="8">Resvd</td></tr>
<tr><td colspan="32">Buffer/Message Size</td></tr>
<tr><td colspan="32">Partner ID</td></tr>
</table>

Fig. 1 The structure of the 128 bit event signatures

comprised of n potentially multi-threaded processes with events of interest happening in these processes. Potential sources of events are the sending and receiving of messages using MPI, collective operations, file I/O operations or the execution of OpenMP-parallel regions.

While it can be configured to write traces to a log file as well, IPM's strength lies in its profiling mode where the time stamps of individual events are not of interest and only statistics of event durations are collected. To efficiently process and store the potentially very large number of events occurring in an application, we derive a unified event signature and encode it as a bit vector. The event signature contains the critical information about the event we are interested in. Figure 1 shows the structure of an event signature bit vector as used by IPM with 128 bits.

- **Event ID:** Corresponds to the numeric encoding of the event being monitored (completion of an `MPI_Send` command, execution of an OpenMP parallel region, etc.). 12 reserved bits correspond to 4,096 different types of supported events.
- **Region ID:** Users can manually mark regions of interest using the `MPI_Pcontrol` mechanism. IPM implements these calls and creates an internal data structure to represent the region. Event statistics are then computed globally (for the entire application) as well as for each marked region separately.
- **Thread ID:** This 8 bit field encodes the ID of the thread in which the monitored event originated. In IPM there is currently only one hash table per process, which is manipulated exclusively in the sequential portions of the application. Multithreaded data is kept in thread-safe data structures local to a module (cf. Sect. 2.3) until a sequential region is reached, at which point the central hash table is updated.
- **Callsite ID:** IPM tries to derive the call site of a monitored function by walking the call stack using `libunwind` or other stack walking libraries. Recording the call site has the benefit of being able to differentiate between MPI calls with the same signature (communication partner, message size, …) but with different dynamic contexts.
- **Partner ID:** This field encodes the communication partner for MPI operations and the file ID for file operations. For collective MPI operations this entry holds the root of the operation and for point to point operations the peer rank is either derived from the arguments to the MPI call or by examining the `MPI_Status` structure.

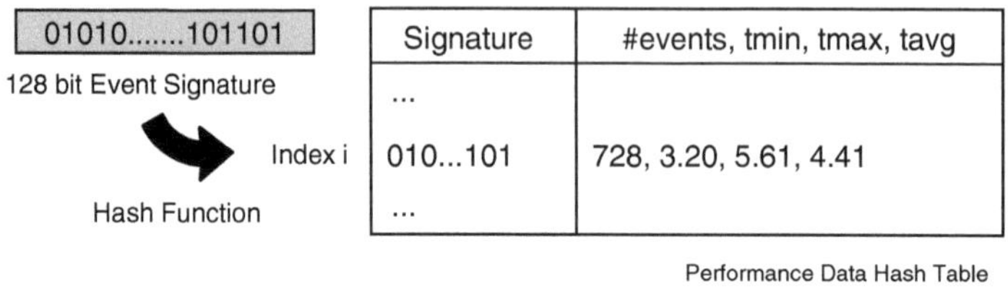

Fig. 2 Event signatures and the hash table used by IPM

- **Buffer/Message Size:** This field encodes the length of the message for MPI operations and the number of bytes read or written for I/O operations. 32 bits are reserved for this field, corresponding to buffer sizes of up to four gigabytes per operation.

IPM observes events at runtime as they are happening in the application, computes their signature and updates the event's statistics in a performance data table. In general, many events may be mapped to the same signature, such as an MPI message exchange between communication partners in a loop with the same parameters. To store and process the performance data efficiently, IPM uses a hash table to implement the performance data table. The event signature is used as the hash key and the hash values are the number of occurrences, the minimum, maximum, and sum of the duration of the events (cf. Fig. 2). Once events from different sources are stored in the hash table, they can be processed and analyzed in a uniform manner.

2.2 *Performance Data Event Sources*

To monitor MPI events, we use the standard PMPI interface. Currently we monitor file I/O at the level of the standard C library calls (`fopen()`, `fclose()`,...) and Unix system calls (`open()`, `close()`,...), by using dynamic library interposition [11] or wrapping these calls at link time for static binaries.

To monitor OpenMP regions we rely on compiler inserted instrumentation as provided by the PGI and Cray compiler suites. If instructed, the compiler will insert instrumentation points in and around OpenMP constructs that are implemented and monitored by IPM. The structure of these calls is largely similar to the POMP calls [9] added by the OPARI source-to-source instrumenter [10]. As an important difference, the source code instrumentation approach requires a recompilation and often reaches its limits when users perform non-standard preprocessing on their source code, such as the inclusion of OpenMP pragmas from header files. The compiler-based instrumentation employed by IPM has no such limitations.

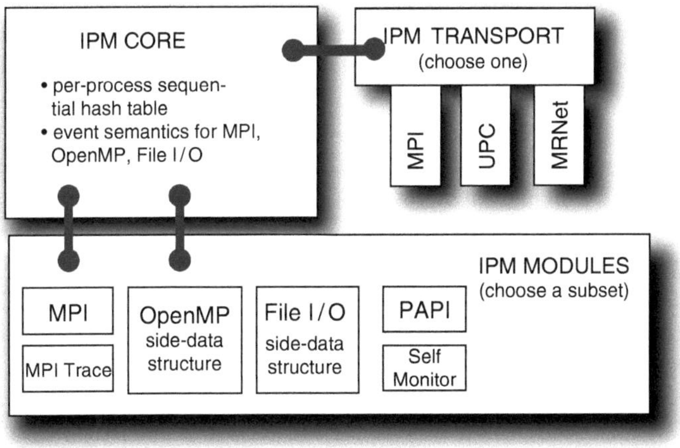

Fig. 3 The modularized design of IPM

2.3 Module Concept

In order to adapt to different requirements and system capabilities, IPM is designed as a modularized system. The schematic system architecture of IPM is shown in Fig. 3. The IPM core module is responsible for initialization of centralized data structures such as the hash table and for registering all available modules. IPM modules are a mechanism of compile-time specialization and not dynamically loadable components, although some runtime configuration is possible using environment variables.

The transport module refers to the mechanism used by IPM to aggregate results across tasks. Since IPM can be used to monitor a parallel application on a per-process level, performance data is co-located with the application processes and needs to be aggregated to compute the IPM application report and log file. Currently the transport module is available for MPI only, but nothing precludes the usage of other communication mechanisms, such as UPC or MRNet [13].

The monitoring components for MPI, OpenMP, and file I/O are each implemented as a separate module. Any subset of these feature modules can be selected to form a valid IPM installation. Notably, the monitoring of file I/O is available for purely sequential jobs without any remaining dependency on MPI (i.e., such an IPM installation can be built on machines without having an MPI compiler and runtime installed). We have used this feature successfully in collaboration with LBNL and CERN scientists analyzing the AthenaMP code [1] which is used for processing data from the ATLAS particle physics experiment.

A number of other modules are currently under construction, such as an MPI-IO component, a module for monitoring the CUDA runtime for GPU enabled systems, and a module for analyzing network interface counters such as those found on

Infiniband hardware. Regular (CPU) hardware performance counters are accessed using a module that encapsulates PAPI [12] functionality. IPM supports component PAPI and reads an environment variable `IPM_HPM` for a user-specified list of counters. Any combination of counter names (from several PAPI components, such as `PAPI_FP_OPS` for floating point operations of and `ETH0_RX_BYTES` for the number of received bytes on the first Ethernet adapter) can be specified and IPM will keep track of which event maps to which PAPI component.

Finally, there is module for self-monitoring of IPM's activity. If enabled, an appendix to IPM's profiling report is available detailing some of the important internal runtime statistics such as hash table fill rates, time in `MPI_Init` and `MPI_Finalize` on IPM's behalf and communication amount and time used for performance data aggregation by IPM.

2.4 Performance Data Output and Post-Processing

IPM's output comes in two forms, each of which can be configured to be disabled, or delivered in a `full` or `terse` format. The most basic output IPM delivers is a banner written to the terminal immediately upon application exit which holds some of the most important high-level job metrics. These include the consumed wallclock time, the number of processes, threads, and nodes the job ran with, and the overall percentage of time spent in MPI, OpenMP parallel regions, and for performing file I/O. Straight forward access to high level metrics is an important aspect of IPM's philosophy as it allows users and computing center staff to get an idea of the overall execution characteristics and resource consumption of an application.

By setting the environment variable `IPM_REPORT` to `full`, a more detailed version of this banner is printed, containing individual events ranked by their contribution. An example showing the full banner of an application is given in Fig. 4. The full banner details the contributions to the wallclock time by MPI and file I/O calls and the time spent in parallel regions as the minimum, maximum, and average over all ranks. The distribution of the time in OpenMP (`OMP`) and the idle time in OpenMP parallel regions (`OMP idle`) at the end of parallel and worksharing regions is displayed in a similar way. The last section lists all individual contributing events sorted by their summed wallclock time. The special event `OMP_PARALLEL` refers to the execution of a parallel region and the `[count]` column refers to the number of executions of this parallel region.

Analyzing the performance of individual processes is possible by requesting an IPM log file. This file in XML format holds detailed information for about each rank's events and includes a full copy of the hash table if `IPM_LOG=full` is specified. The XML file is written sequentially by a designated application process (rank 0 by default) which receives each process' performance information, one by one. At high concurrencies a parallel writing scheme is employed. In this case MPI-IO is used and each rank writes to its portion of the log file in parallel. This

```
##IPM#########################################################
#
# command   : ./a.out
# start     : Sun Mar 14 16:55:39 2010  host      : nid01829
# stop      : Sun Mar 14 17:04:33 2010  wallclock :   533.12
# mpi_tasks : 2048 on 1024 nodes        %comm     :    29.41
# omp_thrds : 6                         %omp      :    50.63
# files     : 12                        %i/o      :    12.09
# mem [GB]  : 2774.44                   gflop/sec :   418.58
#
#           :      [total]        <avg>         min          max
# wallclock :   1091671.57       533.04      532.99       533.12
# MPI       :    321034.43       156.76      109.03       239.23
# I/O       :    131947.08        64.43       11.83       113.87
# OMP       :    552665.28       269.86      205.07       305.36
# OMP idle  :     48262.98        23.57       21.30        27.40
# %wall     :
#   MPI     :                     29.41       20.45        44.88
#   OMP     :                     50.63       38.47        57.28
#   I/O     :                     12.09        2.22        21.36
# #calls    :
#   MPI     :     76235998        37224       37223        37320
# mem [GB]  :      2774.44         1.35        1.35         1.36
#
#                          [time]        [count]         <%wall>
# OMP_PARALLEL          552665.28      131439989           50.63
# MPI_Allreduce         247648.04       14438400           22.69
# fread                  69813.27        5488640            6.40
# ...
#
##############################################################
```

Fig. 4 An example full application banner as delivered by IPM

mechanism is very efficient and scales well to the full size of machines. For example, we have observed that writing the full IPM log at 72,000 processes takes less than 2 min on the Cray XT5 'Kraken' at Oak Ridge National Laboratory.

2.5 HTML Profiling Report

The IPM XML file is input to a parser that generates an HTML representation of the profiling report. Using standard HTML to visualize performance data has the advantage that no special graphical user interface (GUI) is required to view the data. A new parser and techniques using advanced HTML/Javascript charting to address the scalable visualization of data at high concurrencies are currently under development.

Among others, the HTML profiling page contains these entries:

- The data contained in the text-based banner is reproduced in a table at the top of the profiling report.

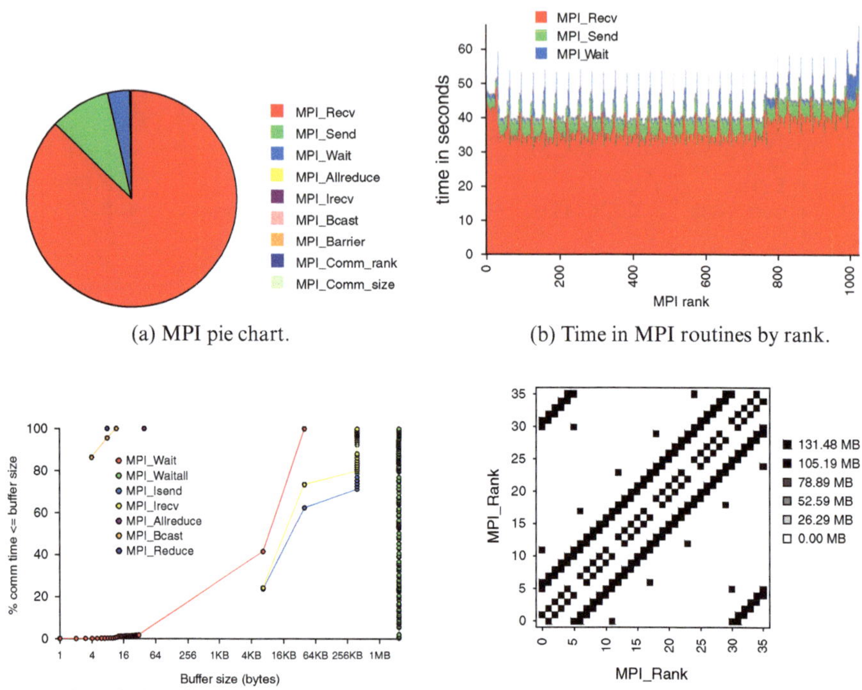

(a) MPI pie chart.

(b) Time in MPI routines by rank.

(c) Cumulative distribution graph of buffer sizes.

(d) Communication topology graph.

Fig. 5 Some of the performance data displays provided by IPM

- A pie chart (Fig. 5a) displays the breakdown of the total MPI time into the various contributing MPI calls such as `MPI_Allreduce` or `MPI_Wait`.
- For each monitored hardware counter event, the minimum, maximum and average values are displayed as well as the location (rank) of where the minimum and maximum values are achieved.
- A load balance line graph showing the consumed DRAM memory, floating point rate, and wallclock time. The horizontal axis is the rank dimension and the graphs are available both in sorted (by memory, FLOPS, wallclock time), as well as unsorted (natural rank order) variant.
- A stacked load balance graph shows the breakdown of the MPI time into individual MPI calls over the rank dimension. An example for this graph (sorted by rank) is shown in Fig. 5b. This type of display is especially useful to identify and locate load imbalance situations.
- Cumulative distribution graphs provide an understanding of the message size distribution of an application. The horizontal axis is the buffer size n used in the operation and the vertical axis denotes how many calls have had a buffer size smaller or equal to n.

- A similar cumulative distribution graph is also provided where the accumulation is not performed over the number of calls but the time spent in the messaging operation instead. The graph in Fig. 5c shows an example.
- A communication topology graph as shown in Fig. 5d. This graph shows the amount of data exchanged between a pair of processes. The sending process is depicted on the horizontal axis, the receiving process is shown on the vertical axis.

3 Efficiency of IPM

In this section we summarize the results we achieved from three sets of experiments we conducted to assess the efficiency of IPM. For details please consult [3].

In the first experiment we tried to asses the raw hash table performance of IPM. 10 million events are prepared with a varying number of unique keys (corresponding to the horizontal axis) and stored in the hash table with 32K entries, including updating the counts and timing statistics, but not actually getting time stamps. The performance depends heavily on the compiler and the optimization flag used but with reasonable settings for production code IPM can process the 10 Mio events in less than 0.8 sec. Performance depends on the number of unique keys used, as collisions become more likely when the hash table fill rate approaches 100%. In the vast number of application test cases we have observed with a 32K hash table, the fill rate is well below 50%.

In the second experiment, we assess the contribution of acquiring time stamps as part of the monitoring process. Here we again record 10 million events, this time by calling `MPI_Comm_rank()` in a simple MPI program. The results show that IPM can handle at least roughly 10 million events per second per application process on a contemporary platform. This number should exceed the messaging rate of an individual process in today's HPC applications.

Finally, we studied the effect IPM monitoring has on a full application by conducting an ensemble study of the MAESTRO application with and without IPM enabled. We find that on average IPM causes an overhead of less than 0.25% for the application kernel of MAESTRO and less than 0.5% for the whole application (including writing the profiling report). On most HPC resources today these perturbations are indistinguishable from natural runtime variations observed due to contention for shared resources.

4 Related Work

There are a number of related performance analysis tools for scientific MPI and OpenMP applications. TAU [8, 14] is a versatile tool set for MPI, OpenMP and Pthreads codes using source instrumentation for function level data collection.

HPCToolkit [17,18] is a tool based on binary analysis and statistical sampling which has its strengths in the detection of code hotspots but places less focus on analysis of communication behavior. Scalasca [5, 16] is a tool for tracing of MPI and OpenMP applications based on a replay of the communication events and the identification of common patterns of inefficiency. Runtime summarization (profiling) capability was added recently. In addition to tools from academic and research labs, vendor-provided tools such as Cray PAT [2] and Intel Trace Collector, Trace Analyzer, and Thread Profiler [6, 7] are available but frequently limited to their particular platforms. ompP [4] is full-featured OpenMP profiling tool that relies on source code instrumentation using the Opari [10] source-to-source instrumenter. In the case of IPM the limitations and shortcomings of source-to-source instrumentation are not acceptable, and as a consequence IPM uses compiler inserted instrumentation to gather high level OpenMP metrics.

In contrast to the tools mentioned above, IPM is designed to give a high level overview of application activity while introducing minimal overheads and being usable in a "flip of the switch" fashion. Given a compatible setup, enabling IPM can be accomplished with a single command such as `module load IPM`. This usage scenario precludes the usage of source code instrumentation (TAU), trace replay (e.g., Scalasca), and binary analysis (e.g., HPCToolkit).

5 Conclusions

We have presented IPM, a modularized application performance and workload analysis tool for MPI, OpenMP and file I/O. Compared to the old implementation of IPM, which was focused only on MPI, we have extended the monitoring coverage into the areas of most importance for today's supercomputers (threading and file I/O). To the best of our knowledge IPM is the first research tool to utilize the compiler-inserted instrumentation for OpenMP monitoring. Previous solutions based on source-code instrumentation were not suitable for IPM's use case that mandates minimal user involvement.

IPM focuses on ease of use and scales today to several tens of thousands of processors. With our focus on lightweight, un-intrusive measurement of performance we expect to be able to scale to even larger systems with minimal effort.

Our new modular design has allowed us to add measurement capabilities for OpenMP and I/O. It will also allow us to easily add measurement capabilities for new hardware features such as GPUs or other co-processors.

IPM's ease of use has enabled the NERSC center to collect over 310,000 profiles of jobs running 20 min or longer on three of its systems over the past six years to gain valuable insights into user and application behavior. Likewise, users benefit from the usage of IPM in their day to day code development and production runs by developing an understanding of resource requirements and tradeoffs. This is critical when dealing with varying "performance weather" on systems or when porting from one machine to another. Often an IPM profiling report can be the starting point for a more in depth evaluation with a tracing or statistical sampling tool.

In the transition to petascale computing, HPC will see shifts in the methods, libraries, and languages deployed by applications. IPM provides a modular way to adapt to the technology shifts in a way that reveals bottlenecks across all the layers involved. We hope that this vertical approach to performance analysis will provide an extensible and durable method to make HPC codes run faster. We plan to work on the addition of monitoring modules for a variety of data sources such as MPI-IO and network interface card counters. We are also currently working on a new parser and data presentation component targeted at supporting high-concurrency data.

Acknowledgements This work was supported by the Bavaria-California Technology Center (BaCaTec) throughout the project "Performance and Workload Characterization for Multi-Core Supercomputers" and by the NSF under award OCI-0721397. This research was also supported by an allocation of advanced computing resources provided by the National Science Foundation. The computations were performed on Kraken (a Cray XT5) at the National Institute for Computational Sciences.

References

1. Binet, S., Winklmeyer, F., Wiedenmann, W., Calafiura, P., Snyder, S.: Harnessing multicores: Strategies and implementations in ATLAS. In: Proceedings of the 17th International Conference on Computing in High Energy and Nuclear Physics (CHEP 2009), Prague, Czech Republic (2009)
2. Using Cray performance analysis tools. http://docs.cray.com/books/S-2376-41/S-2376-41.pdf.
3. Fuerlinger, K., Wright, N.J., Skinner, D.: Effective performance measurement at petascale using ipm. In: Proceedings of The Sixteenth IEEE International Conference on Parallel and Distributed Systems (ICPADS 2010), Shanghai, China, December (2010)
4. Fürlinger, K., Gerndt, M. ompP: A profiling tool for OpenMP. In: Proceedings of the First International Workshop on OpenMP (IWOMP 2005), Eugene, Oregon, USA, May (2005)
5. Geimer, M., Wolf, F., Wylie, B.J.N., Mohr, B.: Scalable parallel trace-based performance analysis. In: Proceedings of the 13th European PVM/MPI Users' Group Meeting on Recent Advances in Parallel Virtual Machine and Message Passing Interface (EuroPVM/MPI 2006), pp. 303–312. Bonn, Germany (2006)
6. Intel Thread Profiler http://www.intel.com/software/products/threading/tp/.
7. Intel Trace Analyzer http://www.intel.com/software/products/cluster/tanalyzer/.
8. Allen, D.M., Sameer, S.S.: Performance technology for complex parallel and distributed systems. pp. 37–46 (2000)
9. Mohr, B., Malony, A.D., Hoppe, H.-C., Schlimbach, F., Haab, G., Hoeflinger, J., Shah, S.: A performance monitoring interface for OpenMP. In: Proceedings of the Fourth Workshop on OpenMP (EWOMP 2002), Rome, Italy September (2002)
10. Mohr, B., Malony, A.D., Shende, S.S., Wolf, F.: Towards a performance tool interface for OpenMP: An approach based on directive rewriting. In: Proceedings of the Third Workshop on OpenMP (EWOMP'01), September (2001)
11. Nakhimovsky, G.: Debugging and performance tuning with library interposers, July 2001. http://developers.sun.com/solaris/articles/lib_interposers.html.
12. PAPI web page: http://icl.cs.utk.edu/papi/.

13. Roth, P.C., Arnold, D.C., Miller, B.P. MRNet: A software-based multicast/reduction network for scalable tools. In: Proceedings of the 2003 Conference on Supercomputing (SC 2003), Phoenix, Arizona, USA, November (2003)
14. Shende, S.S., Malony, A.D.: The TAU parallel performance system. International Journal of High Performance Computing Applications, ACTS Collection Special Issue (2005)
15. Skinner, D.: Integrated Performance Monitoring: A portable profiling infrastructure for parallel applications. In: Proceedings ISC2005: International Supercomputing Conference, Heidelberg, Germany (2005)
16. Szebenyi, Z., Wylie, B.J.N., Wolf, F.: Scalasca parallel performance analyses of PEPC. In: Proceedings of the Workshop on Productivity and Performance (PROPER 2008) at EuroPar 2008, Las Palmas de Gran Canaria, Spain (2008)
17. Tallent, N.R., Mellor-Crummey, J., Adhianto, L., Fagan, M.W., Krentel, M.: Diagnosing performance bottlenecks in emerging petascale applications. In: SC '09: Proceedings of the Conference on High Performance Computing Networking, Storage and Analysis, pages 1–11, New York, NY, USA, ACM (2009)
18. Tallent, N.R., Mellor-Crummey, J.M.: Effective performance measurement and analysis of multithreaded applications. SIGPLAN Not. **44**(4), 229–240 (2009)
19. The Top 500 Supercomputer Sites, web page: http://www.top500.org.
20. Wright, N.J., Pfeiffer, W., Snavely, A.: Characterizing parallel scaling of scientific applications using IPM. In: The 10th LCI International Conference on High-Performance Clustered Computing, March 10–12 (2009)

The J-Cell Project: Developing a Scalable and Robust Runtime Environment for Heterogeneous Many-Core Processors

Thomas Fuhrmann

Abstract J-Cell aims at providing a single system image (SSI) in clusters of heterogeneous many-core processors. The project's fully decentralized runtime environment is designed to hide the underlying hardware complexity, provide robustness in face of hardware failures, and allow applications to scale across a large number of processor cores.

J-Cell is built around the concept of globally accessible objects (GAOs), which can float freely in the system, and which encapsulate all the application state, including the thread's execution contexts. A transaction-based consistency mechanism hides the synchronization latencies. The transaction system's versioning approach also provides the redundancy that guards against failing processor nodes.

First results from simulation studies are encouraging. An actual implementation of the entire system is ongoing work.

1 Introduction

The advent of heterogeneous many-core processors such as the IBM Cell processor has burdened software developers with new challenges. In general, only handcrafted libraries can exploit the performance of these highly complex processor architectures. Software that has been implemented with general purpose languages such as Java usually cannot exploit the processors' parallelism.

The J-Cell project seeks to alleviate that burden with a lightweight runtime environment that hides the complexity of heterogeneous many-core processors from the software developers. The project adopts technologies from the field of distributed computing to provide a *single system image* (SSI) across the cores in a processor, and across the processors in a compute cluster. It particularly aims at

T. Fuhrmann (✉)
Technische Universität München, Boltzmannstr. 3, 85748 Garching, Germany
e-mail: fuhrmann@in.tum.de

C. Bischof (eds.), *Competence in High Performance Computing 2010*,

DOI 10.1007/978-3-642-24025-6_3,

heterogeneous architectures, where the different cores are not equally well suited for all kinds of tasks. J-Cell's SSI hides the hardware details so that applications become independent from the actual architecture.

The project's key technology is a fully decentralized runtime environment that handles networks of globally accessible objects (GAOs). Each GAO can represent a piece of code, an execution context, or some application data. Immutable GAOs can be replicated for faster access and increased reliability. Mutable GAOs are implemented as chains of GAO versions, where each GAO version is again an immutable data structure.

For its prototype, J-Cell uses the *Java Virtual Machine* (JVM) model as starting point, i.e. the prototype can execute bytecode that has been compiled for the JVM and transcoded to the J-Cell virtual machine. The project also develops a library that provides its GAO technology to applications that have been implemented in C/C++.

The project's research agenda comprises several aspects that deal with the relation of GAOs and GAO versions, for example, object location and retrieval, fast consistency algorithms, self-organized redundancy control, and check-pointing for rollback in case of hardware failures. As side topics, the project also addresses aspects of programming language support and novel architectures for virtual machines. The project does, however, not address issues of automatic parallelization. We require the programmers to explicitly provide enough parallelism in their algorithms.

The project primarily aims at *irregular applications* in high-performance computing (HPC). Unlike many typical HPC applications such as fluid mechanics or particle dynamics simulations, irregular applications have a much more complex memory layout that often requires the processor to dereference pointers. Thus, there is no generic way to map an algorithm's working set into the limited memory of an individual processor core. Still, we assume the typical HPC *batch mode* operation, where an application receives its input only upon start-up.

The project's default example are similarity searches for proteins and ligands in the pharmaceutical industry. In particular, we adopt a commercial feature tree algorithm[1] [11] so that it can efficiently exploit the upcoming heterogeneous many-core architectures. Other potential applications comprise event analysis software in accelerator physics, network simulations in computer science, and all kinds of web-based workflows both, for scientific and business applications. The J-Cell runtime environment shall allow a very large number of application instances to efficiently share the same hardware. Each instance shall be able to dynamically grow or shrink according to its varying demands, and shall be shielded from hardware failures in the compute cluster.

In principle, the resulting J-Cell runtime system should provide *ultimate scalability*, i.e. it should not impose any a-priori limits on the scalability of the system. In practice, the scalability of the applications will of course be limited by the

[1] http://www.biosolveit.de/.

applications' underlying algorithms, for example, their memory access patterns. Those algorithmic aspects are, however, beyond the scope of the project.

This paper gives a brief introduction into the J-Cell project. It presents an overview of both, first results and ongoing work from the project. It omits all technical details and performance comparisons, which are discussed in detail in the respectively referenced publications.

The paper is structured as follows: Section 2 gives an overview of the J-Cell system architecture. Section 3 describes the object location and retrieval system. Section 4 briefly describes the DecentSTM protocol, which is used in the transaction-based consistency mechanism. Section 5 explains how we exploit the redundancy that DecentSTM introduces to ensure robustness in face of hardware failures. Section 6 briefly compares J-Cell to similar projects. Section 7 describes our ongoing and future work. Section 8 concludes the paper with a brief summary of J-Cell's key research goals.

2 System Overview

J-Cell's core principle is the concept of objects as universal state containers. These objects represent not only application-level data, but also execution contexts and code. They form a directed graph with object references as edges. The actual location of an object is not important for the functioning of the application, provided that the runtime can resolve the reference at runtime (see Sect. 3). (The object placement is of course important with respect to performance, see Sect. 7.)

Each application begins with one object – the application's so-called *primordial execution context* (PEC). The PEC contains the application's parameters. It also references the code of the application. In general, this piece of code references further pieces of code, namely the libraries that the application binds to. If the application spawns a new thread, these references to the (library) code are copied to the new execution context. Each execution context's stack contains references to (some of) the objects that the application has created. These objects can again contain references to other objects. As a result, a running application is entirely represented by the resulting object graph, which is rooted in the application's PEC.

The objects in this reference graph are not bound to a physical memory location. Any processing component that has obtained a reference to such an object can read and write that object. In J-Cell, each such GAO is a chain of immutable GAO versions. Whenever a thread's object modifications have to become visible to the application's other threads, i.e. whenever the thread encounters a *memory barrier*, the runtime system creates a new GAO version containing these modifications.

Memory barriers are prescribed by the memory model of the respectively used programming language. In J-Cell, we chose a transactional model in the context of Java. Other memory and consistency models might also be feasible. An according comparison is ongoing work in the project (see Sect. 7).

In J-Cell's software transactional memory (STM) paradigm, all code is executed in transactions. This means that each thread is a chain of transactions. Each transaction operates on local copies of the GAOs, only. Thus, a transaction does not directly modify the GAOs. Upon commit, the runtime system determines if any conflicts have occurred, e.g.if two transactions have modified the same GAO. If so, all but one of the conflicting transactions must roll back. They can do so very easily, because the GAOs have not yet been modified and the runtime system thus only has to discard the (modified) local copies. The local copies of the succeeding transactions, however, become the new head versions of the respective GAOs. Thus, a GAO seemingly migrates to the processing node that has most recently committed an update for that GAO.

Unlike most other STM systems, J-Cell does not require a central agent to resolve potential conflicts. Rather, upon commit, all modified objects become tentative versions of the respective GAOs. Program execution can proceed while a randomized consensus protocol handles concurrent commits (see Sect. 4). Thereby, the system hides the synchronization latencies at the expense of a potential rollback, if a conflict cannot be resolved.

The following sections describe the mentioned J-Cell technologies in more detail.

3 Object Location and Retrieval

Current processor architectures such as IBM's Cell processor or Intel's single-chip cloud computer (SCC) indicate that future processors will have to give up cache coherence between the processor's cores. Both, the Cell and the SCC give up the illusion of a large, flat memory space. They rather treat the core's fast on-chip memory as individual memory regions. The Cell – and partly also the SCC – use the fast on-chip memory as message passing buffers, which are not part of the slow off-chip main memory address space. As a result, the software must explicitly manage the on-chip memory.

J-Cell aims at hiding this complexity. Our runtime system decouples object references from the objects' storage location [1]. It keeps track of the references so that it can resolve them regardless of the actual object location. In particular, *object references are no pointers*.

The project has studied various different approaches to efficiently and quickly resolve object references [12]. Meanwhile, it has settled for a proxy-based indirection mechanism that keeps track of relocated objects.

All objects in a node's main memory contain only references to the node's *indirection layer*. This indirection layer can then resolve the referenced object's actual location: If the referenced object is stored locally, the indirection layer can resolve the reference to a pointer in the local address space. Otherwise, the node has to communicate with another node to retrieve a copy of that object. The indirection layer denotes which node should be queried for the object in question.

In some cases such a request must be resolved recursively: When an object is relocated, the node at the most recent storage location creates a proxy in its indirection layer so that requests for that object can be forwarded accordingly. The replies for such requests reveal the object's current location so that the nodes can update their indirection layer entries. After a while, the proxies are cleaned-up by the system's garbage collection mechanism. Thereby, the system avoids indefinitely growing chains of proxies.

This approach avoids global coordination among the nodes. Each location update involves only two nodes: the object's previous and new host. Since in J-Cell, location updates mostly occur with STM commits, the location update protocol is part of the consensus protocol. Hence, object relocation does not impose any additional communication overhead.

First performance analyses show that the proxies do not significantly add to the access latency. In a simple red-black tree application, we found a proxy indirection overhead of about 5%. A proactive notification protocol can avoid that overhead, but we found it to increase the communication overhead by a factor of 2.6. Thus, it depends on the hardware architecture if those proactive notifications are worth the effort. If the communication bandwidth suffices, such a protocol might not affect the applications.

Before we move on to the next component in the J-Cell architecture, the fully decentralized consensus protocol, we want to point out that the said access latency overhead is caused by the distributed nature of the underlying hardware system. Thus, we cannot avoid it entirely. Moreover, within a transaction, the runtime works in fast on-chip memory without further using the described indirection mechanism. Thus, the runtime system's just-in-time (JIT) compiler can replace the said object references with locally valid pointers. (The implementation of such a JIT is not part of the J-Cell project.)

4 Fully Decentralized Consensus

Consistency in distributed systems requires communication. Traditional, lock-based approaches cause a high latency, because a thread must pause until it has successfully obtained the lock. STM allows a thread to proceed optimistically until the end of the transaction. Only then, the system has to resolve the potential conflicts.

As part of the J-Cell project, we have developed a fully decentralized randomized consensus protocol, the DecentSTM protocol [2]. It treats all objects as versioned entities. Each GAO has a so-called *version history list*, which contains the GAO's head version and a limited number of outdated object versions. The latter can avoid read conflicts, and thereby – in some cases – keep a transaction from failing. Keeping outdated object versions also improves the system's redundancy (see Sect. 5).

Upon commit of a transaction, the modified object content tentatively becomes (one of) the new object version(s). In general, each object can have several tentative object versions. They form the so-called *tentative version tree*. Different branches in that tree reflect the GAO's conflicting versions. The consensus protocol ensures that all involved nodes settle for a consistent subset of transactions to succeed.

Since the consensus protocol is randomized, it can be biased by, for example, considering the GAO versions' value, as determined by the time and energy that their calculation required. Thereby, DecentSTM can prefer to roll back cheap transactions, rather than the costly ones. When the consensus protocol has settled and global consensus has been reached, the succeeding version is moved into the GAO's globally consistent version history list. (The integration and analysis of such a bias is ongoing work.)

The DecentSTM protocol is not only a fully decentralized protocol; it also decouples the transactions from the consensus phases: A transaction can proceed optimistically and read a tentative GAO version, at the risk of having to roll back, if the consensus protocol happens to reject the chosen version. (This is a third type of conflict besides read and write conflicts, which also occur in other STM systems.) Assuming that in the general case the runtime system picks a version that will succeed, the system can increase the throughput and avoid the latency of the consensus protocol.

Even though, DecentSTM has been designed for transactions, its inner workings also apply to non-STM systems, because, in essence, DecentSTM only regulates when thread-local modifications become globally visible. When we apply, for example, the standard Java memory model, we do not need the consensus protocol, because there are no transactions to fail. Upon reaching a memory barrier, the thread-local object modifications simply become the objects' new versions. Global consistency has to be enforced at application level by using locks. We are currently investigating ways to hide the communication latency in the lock-based consistency paradigm, too (see Sect. 7).

5 Redundancy

DecentSTM inherently produces snapshots of the system's state, because each transaction must be able to roll back if it does not succeed in the consensus protocol. These snapshots come in form of the GAOs' version history lists. If these lists span multiple physical nodes in the system, they also provide redundancy that allows the system to resume after a hardware failure.

J-Cell's redundancy control mechanism aims at fostering such a preferable situation: For each GAO whose versions are stored on at least two different processor nodes, the protocol keeps (at least) two versions on different processor nodes. But since keeping arbitrary versions does not guarantee a consistent snapshot, the redundancy control mechanism has to consider the dependencies between the transactions, especially between the GAO versions in the read and write sets. Roughly speaking, the redundancy control mechanism has to ensure that the GAO versions that a transaction has written are stored on another node than the GAO versions that it has read. This condition ensures that if the GAO versions that a transaction has written are lost due to a hardware failure, the transaction can be repeated on another node using the very same GAO versions that the transaction read during the first execution.

Let us describe that mechanism in more detail: The dependencies between the written and read GAO versions induce the so-called *transaction dependency graph* (TDG). Each node in the TDG contains all the GAO versions that the respective (successful) transaction has committed, references to the GAO versions that the transaction has read, as well as the execution context of the respective thread at the time when the transaction has committed. Hence, these nodes in the TDG are called *comprehensive transaction records* (CTR).

The TDG is a directed, acyclic graph. When a transaction commits, it adds a CTR to the TDG. This CTR has edges to all the CTRs from which the transaction has read GAOs. (Written GAOs are also included, because we emulate a read operation before each write operation.) As long as a CTR still contains GAO versions that have not yet been replaced by a more recent version, it is called *current* CTR. So-called *past* CTRs contain only GAO versions that are not required any more, but only contribute to the system's redundancy.

Losing a past CTR does not affect the application. It only reduces the available amount of redundancy. But when a current CTR is lost, the application cannot proceed. (It could, if the application happened to not read any of the therein contained GAOs.) To enable the application to resume, the lost transaction is repeated so that the lost CTR is re-created. The transaction can exactly be repeated, because the TDG has recorded the exact dependencies of the respective CTR.

J-Cell's redundancy control mechanism ensures that all the required CTRs are stored on a different node. Moreover, it stores the respective dependency relations redundantly, so that they can be retrieved. Thus, assuming only independent failures, the system can recover with high probability.

A simple collapsing rule determines which of the past CTRs may be combined without violating the said redundancy rule. Combining two CTRs (on the same node), allows to eliminate superfluous redundancy. Consider, for example, a chain of dependent CTRs: A depends on B, and B depends on C. If A and B happen to be stored on the same node, the system may combine A and B and thereby eliminate all GAO versions from B that have been updated in A. The system will, however, keep C, because it provides redundancy in case $A \cup B$ is lost. Any further CTR that C might depend on can be discarded (provided no other CTR depends on it), because we assume that only either A or C can be lost, but not both. (The details of that redundancy mechanism are beyond the scope of this paper, see [9].)

6 Related Work

There are several Java runtime systems that provide a distributed shared heap. Probably most widely known is the Jessica system [4]. It allows objects to be replicated across several compute nodes. Object modifications are synchronized via the objects' home nodes. Unlike Jessica, J-Cell aims at a fully decentralized system, because such a system can provide better robustness and scalability.

Another well-known distributed Java system is JavaParty [5]. It supports remote object access with the help of an efficient remote method invocation (RMI) system. Unlike JavaParty, J-Cell integrates deeply into the runtime system. Thereby, it can support the modern many-core architectures, which cannot execute off-the-shelf Java VMs efficiently.

Other authors modify existing VMs to support a distributed shared heap and thread migration. Quitadamo et al. [10] extend the *JikesRVM* with a thread serialization mechanism, which allows them to migrate threads at runtime. In contrast, J-Cell pursues a more thorough approach that treats data and execution contexts as freely floating entities. Again, this allows J-Cell to efficiently run on modern many-core architectures.

Liu et al. [6] develop *Fabric*, a platform for secure distributed computing. Like J-Cell, they build their platform on mutual state records and transactions; but unlike J-Cell, Fabric targets cloud computing systems. Thus, they focus more on overlay and security aspects than on the efficient use of many-core processors.

Noll et al. have presented CellVM [8]. McIlroy et al. have presented HeraVM [7]. Both VMs specifically target single instances of the IBM Cell processor. In contrast, J-Cell aims at scalability across a large number of such processors. Moreover, J-Cell seeks to provide robustness against hardware failures.

Recently, two commercial systems have gained widespread interest: Terracotta[2] provides a distributed shared heap, which allows Java applications to better scale in large business systems. Unlike J-Cell, Terracotta weaves additional code into the applications that allows them to access the remote parts of the heap [3]. Again, such an approach is not feasible for us, because of-the-shelf VMs do not fit the J-Cell scenario. The Zing platform by Azul Systems[3] also aims at scaling Java enterprise applications to very large machines. Unlike Terracotta, Azul uses its own runtime system to create a SSI across the processors in a compute cluster. I.e. Azul pursues a similar approach as J-Cell; but being a commercial product, little is known about the Azul VM's inner workings.

7 Future Work

J-Cell is an ongoing project. The current architecture has various knobs to further optimize its performance and energy efficiency. The DecentSTM protocol, for example, allows the system to bias which transaction is more likely to succeed. A useful input parameter for this bias can be the time and energy that a transaction has required. This means the system could prefer the more expensive transactions, rather than those that are cheap to repeat. Other parameters might also be important, especially with respect to ensuring that the application is guaranteed to make

[2]http://www.terracotta.org/.

[3]http://www.azulsystems.com/.

progress. A thorough evaluation of all these parameters is planned, once we have finished implementing our prototype system.

Another such knob in the design is the *object placement strategy*. It is important, because our system cannot hide the read latencies. Thus, ideally, the current GAO versions should be stored close to those threads that will read them in the near future. Most likely, it is impossible to fully control the interaction between the application's control and data flow, the scheduler, and the object placement system; but we expect to be able to develop heuristics that can bias the object placement so that it tends to store the GAO versions in preferable locations.

Closely linked to the object placement strategy is an envisage mechanism that deliberately creates and maintains the system's redundancy. So far, J-Cell only exploits the redundancy that comes with the transaction mechanism. This by-product does not suffice in all cases. If, e.g., an object is never modified after it has been created, there are no additional GAO versions for that GAO. In such a case, our system must explicitly create redundant copies. Such copies can also help to reduce the read latency.

Besides these optimization issues, we also face further design issues. For example, we plan to study solutions that allow applications to have side-effects. Most importantly, we seek to allow input and output operations in the transactions, e.g.network communication. This issue has already been studied in the research community, but we have to see which of these solutions fits best into J-Cell.

Furthermore, we want to study how we can best support applications that use lock-based synchronization. We expect that the fundamental concept of GAOs and GAO versions carries over to lock-based synchronization; but we need to complement our protocols with protocols for obtaining and releasing locks. Most importantly, these protocols need to be robust in face of node failures.

Finally, the project explores further practical aspects with respect to scalable and robust runtime systems. For example, we have to design novel architectures for lightweight virtual machines (VMs) that can run on small processor cores, because current VMs do not fit the modern heterogeneous processor architectures. Moreover, we are looking into the design of protocols that are required for decentralized clouds, because we expect them to become an important means of scientific computing in the near future. Such clouds allow, for example, volunteer computing, i.e. they exploit compute resources that are spread across the Internet. Unlike early attempts such as Seti@Home or Folding@Home future systems must be flexible and allow the execution of any kind of application. We are confident that many insights from the J-Cell project carry over to these architectures as well.

8 Conclusion

In this paper, we have presented the J-Cell project. It aims at providing a SSI in clusters of heterogeneous many-core processors. In particular, J-Cell develops a fully decentralized runtime environment that hides the underlying hardware complexity and provides robustness in face of hardware failures.

J-Cell is built around the concept of GAOs, which encapsulate all the application state, including the thread's execution contexts. Each GAO can come in multiple, immutable versions that form a version history list. If state is lost due to a hardware failure, the runtime can roll back to the most recent consistent version. J-Cell entirely avoids single points of failure, because it only employs fully decentralized algorithms and protocols. J-Cell's transaction-based consistency mechanism also hides the communication latency that is associated with traditional, lock-based consistency mechanisms.

J-Cell is an ongoing project. So far, it has been mostly focused on designing a fully decentralized system architecture. First results about its feasibility are based on simulations, only. An actual implementation of the system is ongoing work.

Acknowledgements The J-Cell project is funded by the Bundesministerium für Bildung und Forschung (BmBF) under grant number 01IH08011. I would like to thank all our project partners (Universität Freiburg and BioSolveIT GmbH) and associated partners (IBM Deutschland Research & Development GmbH, Intel Microprocessor & Programming Research Lab, and MicroDoc GmbH) for their support. I am particularly grateful to the project's PhD students Annette Bieniusa, Benedikt Elser, Bernhard Amann, Björn Saballus, and Stephan-Alexander Posselt.

References

1. Bieniusa, A., Eickhold, J., Fuhrmann, T.: The Architecture of the DecentVM – Towards a Decentralized Virtual Machine for Many-Core Computing. In: Proceedings of the 4th Workshop on Virtual Machines and Intermediate Languages (VMIL) (2010)
2. Bieniusa, A., Fuhrmann, T.: Consistency in hindsight: A fully decentralized STM algorithm. In Proceedings of the 24th IEEE International Symposium on Parallel & Distributed Processing (IPDPS) (2010)
3. Bonér, J., Kuleshov, E.: Clustering the Java virtual machine using aspect-oriented programming. In Proceedings of the 6th Intl. Conf. on Aspect-Oriented Software Development (AOSD) (2007)
4. Fang, W., Wang, C.-L., Lau, F.C.M.: On the design of global object space for efficient multithreading Java computing on clusters. Parallel Comput. **29**(11–12), 1563–1587 (2003)
5. Haumacher, B., Moschny, T., Reuter, J., Tichy, W.F.: Transparent distributed threads for Java. In Proceedings of the 17th International Parallel and Distributed Processing Symposium (IPDPS) (2003)
6. Liu, J., George, M.D., Xin Qi, K.V., Waye, L., Myers, A.C.: Fabric: A platform for secure distributed computation and storage. In: Proceedings of the 22nd ACM Symposium on Operating systems principles (SIGOPS), pp. 321–334 (2009)
7. McIlroy, R., Sventek, J.: Hera-JVM: Abstracting processor heterogeneity behind a virtual machine. In Proceedings of 12th Conference on Hot Topics in Operating Systems (HotOS), (2009)
8. Noll, A., Gal, A., Franz, M.: CellVM: A homogeneous virtual machine runtime system for a heterogeneous single-chip multiprocessor. In Proceedings of the Workshop on Cell Systems and Applications, (2008)
9. Posselt, S.-A.: Design of a reliable, fully decentralized software transactional memory protocol. Diplomarbeit, Technische Universität München (2010)

10. Quitadamo, R., Cabri, G., Leonardi, L.: Mobile JikesRVM: A framework to support transparent Java thread migration. Sci. Comput. Program. **70**(2–3), 221–240 (2008)
11. Rarey, M., Dixon, J.S.: Feature trees: a new molecular similarity measure based on tree matching. J. Computer Aided Mol. Des. **12**(5), 471–490 (1998)
12. Saballus, B., Fuhrmann, T.: A decentralized object location and retrieval algorithm for distributed runtime environments. Technical report, Technische Universität München (2010)

Pre-Stack PRO: A High-Performance Seismic Data Processing Solution

Tobias Götz and Alexander Neundorf

Abstract This paper presents Pre-Stack PRO as a commercial HPC application in the field of the oil & gas industry. After a short introduction to seismic data in general, the paper presents the basic ideas behind Pre-Stack PRO and shows the way it helps customers, improve the way they used to work with seismic data. A closer view to the architecture of Pre-Stack PRO reveals the several levels of parallelization within the software as well as the idea of global collaboration using a single dedicated Pre-Stack PRO cluster from all over the world. Finally some of the fundamentals of Pre-Stack PRO are presented, such as the GPI, the 3D visualization engine and the parallel filesystem.

1 Introduction to Seismic Data

Processing of seismic data in many aspects is a challenging task. For a better understanding, this chapter presents a short introduction to seismic data acquisition and its structure.

For the acquisition of marine seismic data, a vessel is equipped with an airgun and several streamers that carry up to several hundreds of receiver units (hydrophones). The airgun shoots acoustic waves into the underground at a constant timing interval. Those waves are reflected and refracted in the subsurface and finally registered by the hydrophones. For a single shot, depending on the distance to the airgun multiple events are registered (see Fig. 1 on page 40).

During the acquisition process, the ship passes the area of interest several times to finally cover the whole survey. The acquired data is then sorted into a regular grid. This is a necessary step, as typically the ship cannot follow absolutely straight

T. Götz (✉) · A. Neundorf
Fraunhofer ITWM, Competence Center High Performance Computing, Fraunhofer-Platz 1, 67663 Kaiserslautern, Germany
e-mail: tobias.goetz@itwm.fraunhofer.de; alexander.neundorf@itwm.fraunhofer.de

C. Bischof (eds.), *Competence in High Performance Computing 2010*,
DOI 10.1007/978-3-642-24025-6_4, © Springer-Verlag Berlin Heidelberg 2012

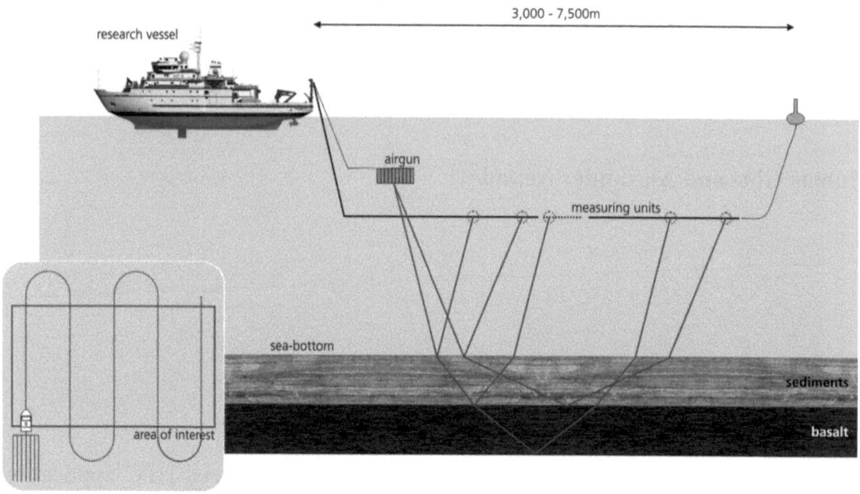

Fig. 1 Schematic representation of the acquisition process: Signals from the airgun are reflected in the subsurface and received by several different hydrophones. The *top* view in the *lower left* shows the travelpath of the vessel over the area of interest

paths due to drifts, winds and other effects. During that process, shots and events are sorted to common depth points (CDP).

The result of the whole process then is a four-dimensional seismic data structure of the underground, where the CDP position given in inline and crossline position represents x- and y-direction, time represents z-direction and the offset represents the fourth dimension. This fourth dimension is the result of having multiple receiver units. For a given point in the subsurface, hydrophones at different distances from that point receive different reflected signals. Those traces are combined together to a gather for that certain CDP (see Fig. 2 on page 41).

First structural analysis and investigations were based on three dimensional datasets where the traces of a single gather were put together into one trace in a process called *Stacking*. Those stacks typically are by a factor of about 100 smaller than the prestack data. But on the other hand it is necessary to extract as much information as possible from the available data to prevent drilling dry which costs tens of millions of Dollars each. Part of this information hides in the gathers, so working on four-dimensional data brings more quality to the resulting stacks and thus reduces risks but also increases the demand for fast processing software, that can handle the big 4D datasets.

2 Introduction to Pre-Stack PRO

Pre-Stack PRO is the first software available, that allows interactive working on large prestack datasets. This Fraunhofer ITWM development is a commercial application since March 2010 and it is distributed via Sharp Reflections, a Fraunhofer

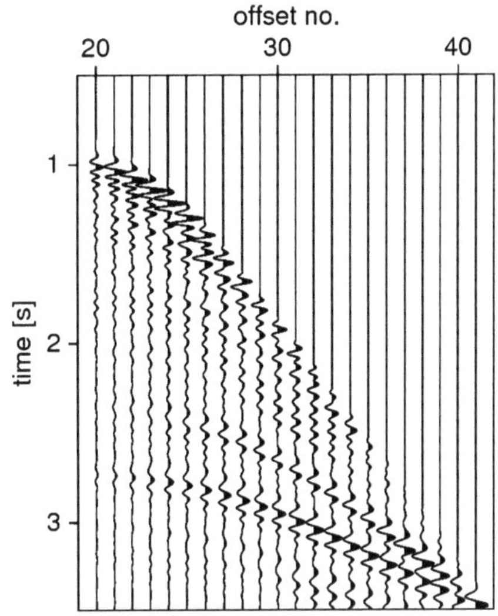

Fig. 2 Seismic gather put together from single traces for different receiver locations. Offset (i.e. distance to the CDP) is increasing from *left* to *right*, so the same event is triggered at a later time

spin-off. As a tool for gather conditioning, visualization and prestack interpretation, Pre-Stack PRO offers a complete set of gather flattening and filtering algorithms, a very powerful 3D visualization engine with realtime volume rendering and a collection of interpretation tools. All based on HPC tools, developed at the Competence Center High-Performance Computing.

By design, Pre-Stack PRO supports global collaboration of geoscience specialists all over the world and reduces the need for sending data and/or travelling, as every user can connect to the Pre-Stack PRO backend on a dedicated cluster using the Pre-Stack PRO client software on his local machine (for more details on the architecture see Sect. 3).

Among the current customers there are global players like Marathon Oil Company (USA), Statoil ASA (Norway) as well as several SME in Europe and Northern America.

2.1 *Gather Flattening*

Flattening the events in a prestack gather is among the most important tasks to improve the quality of the stacks and reduce the risk for drilling decisions. Pre-Stack PRO offers several algorithms for flattening gathers, such as semblance optimization, high-order NMO correction, η-scan correction or continuous velocity analysis tools, each optimized for a certain kind of non-flat event. See Fig. 3 on page 42 to see a given input gather before and after applying Pre-Stack PRO algorithms to the data.

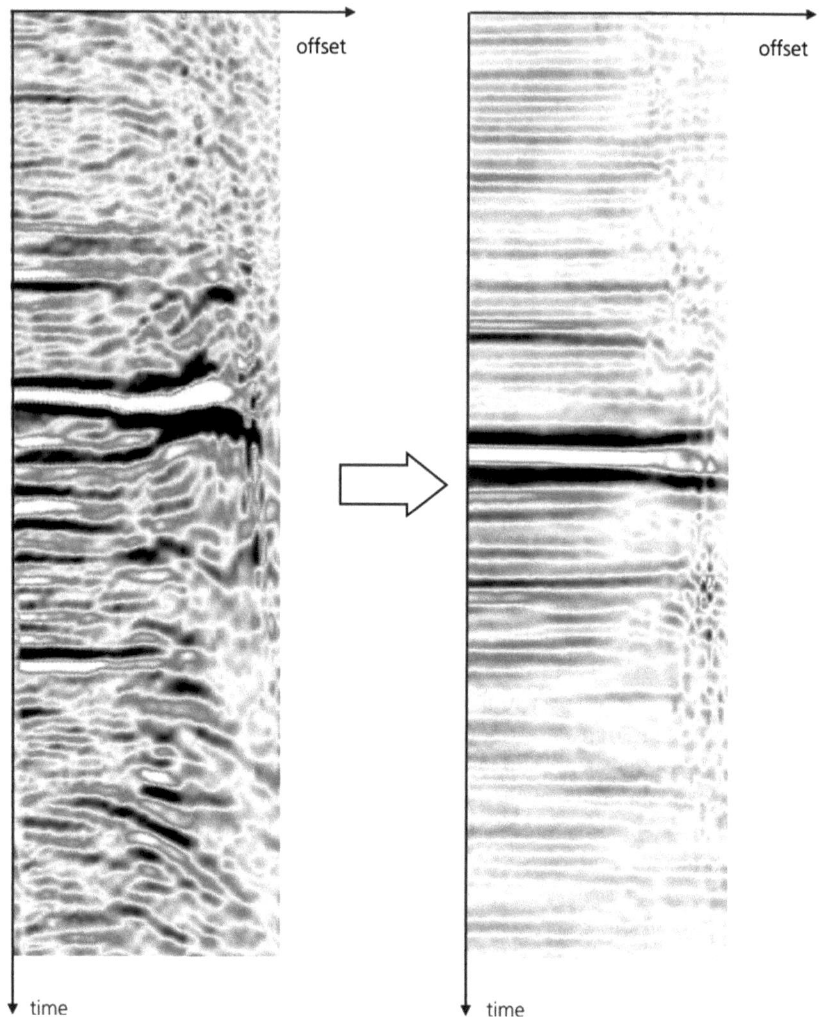

Fig. 3 This figure illustrates the amount of improvement that can be achieved using Pre-Stack PRO on a gather with random noise, multiple reflections and non-flat gathers. After applying several algorithms, the reflectors are very flat and stacking those traces to a master trace will lead to a much better stacked result with more information about the underground

2.2 *Filtering Algorithms*

Besides gather flattening, removing noise and multiple reflections are essential for improving the quality of given prestack data. Pre-Stack PRO offers a high-resolution, high-speed Radon correction for removing noise using the linear Radon and multiples using the parabolic and hyperbolic Radon.

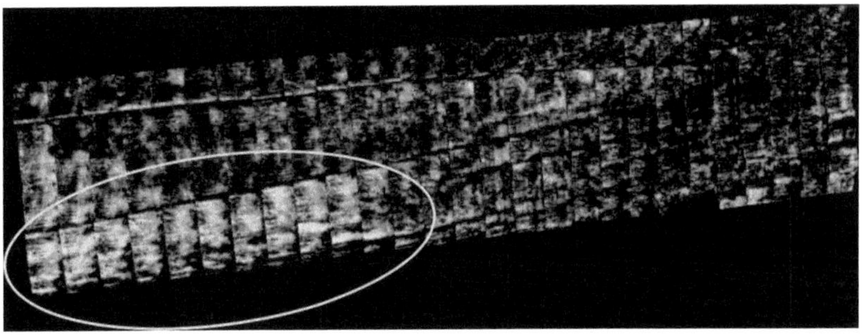

Fig. 4 Amplitude Analysis: In the *lower left* of the volume rendered data, one can see the amplitudes brighten up. This is one indicator for hydrocarbons at this location in the subsurface

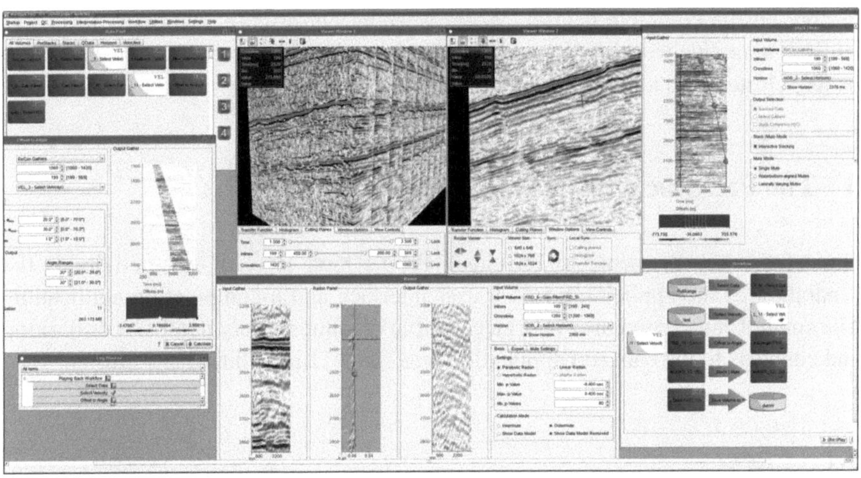

Fig. 5 Screenshot of the Pre-Stack PRO working desktop with datapool, two 3D visualization windows, several algorithm GUIs and the workflow GUI (*top left* to *bottom right*)

An edge preserving, coherency enhancing anisotropic diffusion filter allows filtering of events across the whole dataset in all directions using diffusion equations. This filter removes noise while it preserves and enhances the events in the dataset.

2.3 Amplitude Analysis

Though Pre-Stack PRO has its focus set on processing, there are also some valuable tools for interpreting and analysing prestack data. Figure 4 on page 43 shows a section of a given survey, cut to the area in depth, where the hydrocarbons are. By applying volume rendering to that section, leaving only the high negative amplitudes, the hydrocarbons show up by brightening the amplitudes in the lower left of the image.

2.4 *Working Modes*

By using a partitioned global address space (PGAS) over all nodes in the available cluster, Pre-Stack PRO is designed as an interactive application. Customers who own big Infiniband clusters with several hundred compute nodes, increase the number of nodes Pre-Stack PRO is running on and hence the memory allocated. With an average memory equipment of 32 GB per node, 100 compute nodes allow to allocate 1,600 to 2,400 GB for Pre-Stack PRO PGAS (using 50 to 75 percent of the available memory). This amount will host most of the bigger projects and by increasing the number of compute nodes, the PGAS increases, too.

For the case, there are no more nodes left or customers don't own big clusters, Pre-Stack PRO also supports out-of-core working. Workflows can be recorded with smaller parts of the dataset and after fine-tuning the parameters being replayed out-of-core on the complete dataset. While working out-of-core, Pre-Stack PRO still allows interactive working in parallel. This feature allows usage of Pre-Stack PRO on any cluster with at least two compute nodes.

3 The Pre-Stack PRO Architecture

One of the key elements to Pre-Stack PRO's high level of interactivity is the separation of the application into a user interface and a computation application. The small, Linux based user interface can run on almost every notebook or desktop and connects to the backend application on a (large) Linux Infiniband cluster.

3.1 *Client*

The Pre-Stack PRO client is a Linux and Qt4 based application that can run on almost every Linux system with a good graphics hardware and access to the internet. Client and backend application communicate with generic TCP/IP networking. The client provides the user interface for the full functionality of Pre-Stack PRO as shown in Fig. 5 on page 43.

By design, global collaboration is possible with the client started from any point in the world connecting to a dedicated cluster, where the backend is running and where also the seismic data is stored on a parallel filesystem.

3.2 *Backend*

The Pre-Stack PRO backend is a parallel application for 64bit Linux Infiniband clusters that builds on HPC tools developed by Fraunhofer ITWM's Competence Center HPC. Those techniques are used to set up several levels of parallelization as shown in Fig. 6 on page 90.

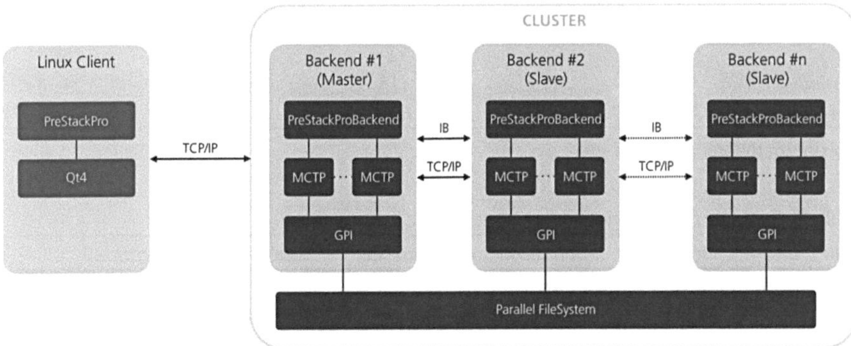

Fig. 6 Pre-Stack PRO architecture with client and backend application showing the levels of parallelization in the backend application

The first parallelization layer is a parallel filesystem to aggregate I/O read and write speed of several (hundred) single hard drives. Pre-Stack PRO can run with every available parallel filesystem, but best performance is achieved using the Fraunhofer Parallel FileSystem (FhGFS, see Sect. 4.3).

On the second layer of parallelization, for the high volume, high frequency data exchange between the single compute nodes in the cluster, the Fraunhofer Global Address Space Programming Interface (GPI) is used (see Sect. 4.1).

While GPI addresses parallelization on the node level, the Fraunhofer Multicore Thread Package (MCTP) is used for the parallelization within a single node. Pre-Stack PRO uses dedicated threadpools for processing, visualization and I/O with one thread per core. Using priority levels guarantees, that threadpools with the most important tasks get more compute time on the CPU. In Pre-Stack PRO the visualization threadpool is favoured above the processing and the I/O threadpools.

Another level of parallelization is added in the optimization step as described in Sect. 3.3 on page 45.

This architecture allows the backend to scale not only with the number of the compute nodes in the cluster, but also with the number of cores per compute node.

In addition to this parallelization, Pre-Stack PRO comes with a powerful CPU-based 3D visualization and render engine (see Sect. 4.2) and highly optimized algorithms for the actual processing of the seismic data.

3.3 *Scalability and Performance*

By the design of the architecture, algorithms in Pre-Stack PRO scale linear with the number of available cores either by increasing the number of cores per node or by increasing the number of compute nodes (see Table 1 on page 46 for an example algorithm).

On top of the overall speed-up by running the algorithms in parallel, Pre-Stack PRO uses highly optimized versions of the original algorithms. Optimization

Table 1 Runtime of Radon with increasing number of compute nodes (cores)

Cores	8	24	52	128	256
Runtime (s)	703	179	113	47	24
Speed-up	–	3.9 (4)[a]	6.2 (6.5)	14.9 (16)	29.3 (32)

[a]Maximum theoretical speed-up

Table 2 Examples for algorithmic optimization

Algorithm	Original	Optimized	Speed-up	Data size	Optimized
2D Median	1.3 sec	0.16 s	8.2	11.4 MB	Computation
Radon	59 sec	2.1 s	28	0.76 MB	Computation
Migration	563 sec	136 s	4.1	224.3 MB	Computation, communication

is achieved with several different techniques like using extended instruction sets (MMX, SSE etc.), optimal memory and cache handling, translating legacy codes from `FortranXX` or other programming languages to current `C/C++` or overlapping communication and computation using the GPI. (See Table 2 on page 46 for an overview of the runtime improvements using those techniques).

4 Fundamentals of Pre-Stack PRO

4.1 GPI

GPI is the Global Address Space Programming Interface, setting up a partitioned global address space (PGAS) over all compute nodes in the backend cluster and Fraunhofer's answer to MPI. While MPI uses message passing for communication and therefore causes CPU load, GPI uses one-sided communication with RDMA instead. To achieve this, fast network connections are needed and GPI supports both, Infiniband and 10 GB Ethernet (Fig. 7).

GPI is threadsafe and supports direct reading and writing to remote memory, both asynchronous and synchronous. GPI also comes with global barriers and global atomic counters. Several different Fraunhofer ITWM projects make use of GPI[1].

4.2 Render- and Visualization Engine

Another feature that distinguishes Pre-Stack PRO from its competitors is the powerful 3D render and visualization engine. Pre-Stack PRO allows simultaneous

[1]www.gpi-site.com.

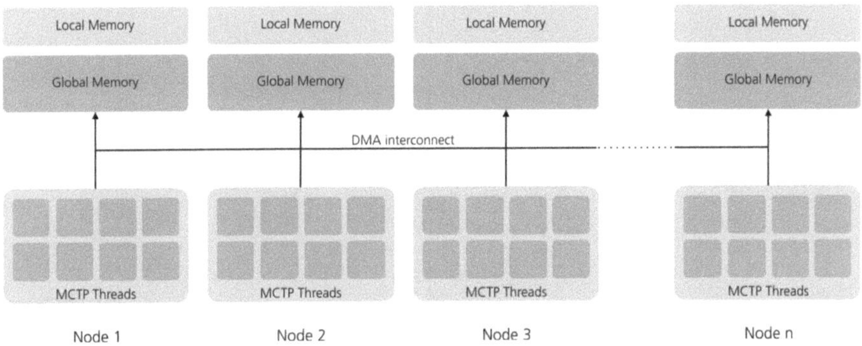

Fig. 7 The GPI architecture

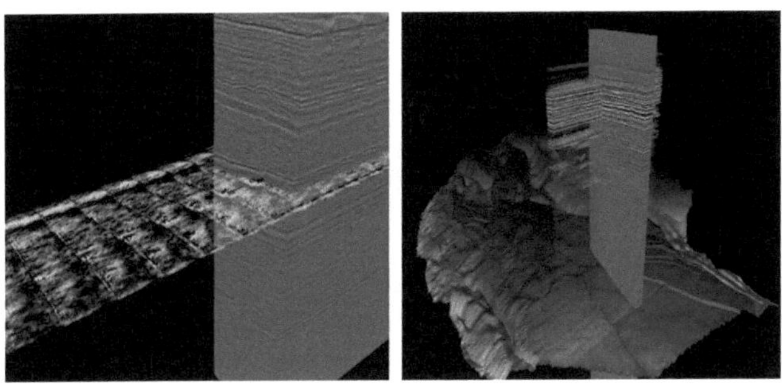

Fig. 8 Screenshots of the Pre-Stack PRO viewer windows using the 3D render engine

visualization of more than 8 GB of seismic data in a single interactive 3D scene and up to four of those viewer called scenes at one time. As the visualization is fully done on CPU, no special graphics hardware on the cluster is required instead only the rendered image is sent back to the client.

Using fast raytracing algorithms, transfer functions can be applied to the seismic data while still working interactively with a framerate of at least 20 fps. Also triangulated surfaces can be displayed together with the voxel based seismic data to define special areas/layers in the subsurface.

Figure 8 on page 47 shows some screenshots of the Pre-Stack PRO viewer windows.

4.3 *Fraunhofer Parallel File System*

The fact that Pre-Stack PRO is able to work on very big datasets using the GPI calls for very fast I/O. Therefore Pre-Stack PRO relies on a fast parallel filesystem to aggregate the I/O speed of the available hard drives (Fig. 9).

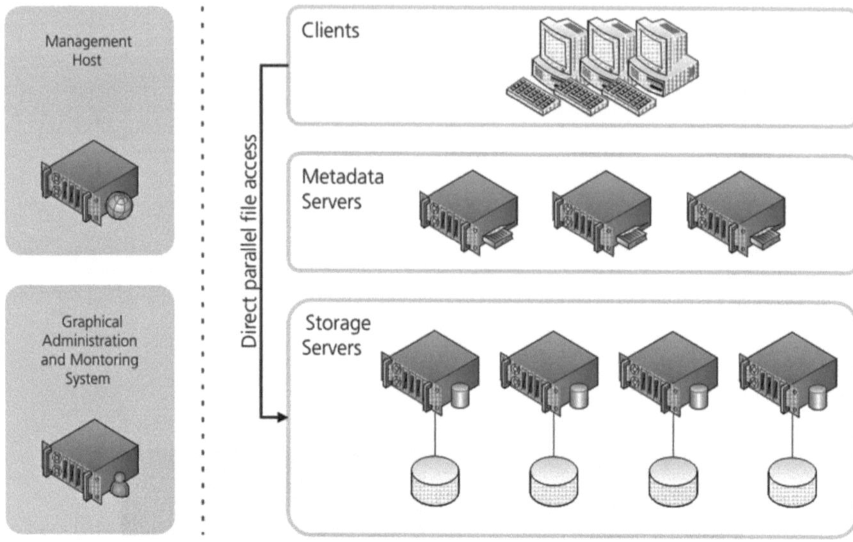

Fig. 9 Structure of the FhGFS with clients, metadata and storage servers

The Fraunhofer Parallel Filesystem (FhGFS[2]) is another Fraunhofer development and optimized for HPC usage. It supports native Infiniband as well as TCP/IP and is designed to scale simultaneously with the compute cluster due to distributed meta data servers and storage servers. This distribution is one of the key elements to the high performance, FhGFS is delivering. Other advantages of the FhGFS:

- As a Kernel module, no patches are needed
- All Kernel versions since 2.6.16 are supported
- It's free to use

5 Summary

Pre-Stack PRO is a powerful processing and gather conditioning tool for the Oil & Gas industry that allows interactive working with up to several Terabyte of seismic prestack data. By splitting the application into a user interface client and a backend compute application, global collaboration of geoscientists is made possible without sending Terabytes of data or travelling.

Having several levels of parallelization is one of the fundamentals of Pre-Stack PRO's interactivity. With the Global Address Space Programming Interface (GPI) allowing direct reading and writing of remote memory reduces load on the CPU

[2]www.fhgfs.com.

introduced by message passing based communication. Dedicated threadpools with priority levels guarantee parallel processing and visualization and the fast parallel filesystem builds the basis for loading the desired data fast from hard disk into the GPI space.

Algorithms that are implemented in Pre-Stack PRO first have to pass an optimization step. Besides speed, also quality of the algorithms is constantly improved and monitored using automated tests.

Finally, the rich visualization tools in Pre-Stack PRO allow a geophysicist a new view into his seismic data. Synchronization between several viewer windows allows to investigate several different volumes at a time, direct links between the 3D viewers and the algorithm GUIs enable the user to scan through the volume and look at gathers at dedicated locations.

Acknowledgements Special thanks to Statoil ASA for providing the seismic data set shown in some of the illustrations in this paper.

A New Scalable Monitoring Tool Using Performance Properties of HPC Systems

Carla Guillen, Wolfram Hesse, and Matthias Brehm

Abstract We present a monitoring and analysis tool prototype for system wide monitoring of High Performance Computers. The tool uses formal specification of properties which are based on hardware counters. These evaluate the performance at different granularities, namely at core, application and partition graininess. The information obtained is aimed at detecting single node performance as well as parallel execution performance. The goal is to identify performance bottlenecks in running applications as well as the general system behaviour. The scalability in our prototype for highly parallel machines is achieved through a distributed software architecture. We use an analysis agent at each partition. These agents communicate to a high level agent using a communication protocol based on TCP/IP. The high level agent has as a main task the synchronisation of the rest of the agents. Moreover, the analysis agents have the capability to use OpenMP within each partition to parallelise their monitoring tasks. Our approach used to tackle the storing of large amounts of information is achieved by data reduction. Only the properties that detect a bottleneck are stored, thus we don't compromise the quality of the needed monitoring information.

1 Introduction

Massive parallel systems are an expensive resource and analysing the performance of its usage is important to detect and correct weaknesses or to foresee possible breakdowns. There are several challenges associated with efficient monitoring of large scale parallel systems. The most indispensable one to tackle is scalability. The two main problems are inefficiency or low performance during measurement and processing and the storage of large quantities of data. In this paper we present

C. Guillen · W. Hesse (✉) · M. Brehm
Leibniz Supercomputing Centre, Boltzmannstr. 1, 85748 Garching, Germany
e-mail: guillen@lrz.de; hesse@lrz.de; brehm@lrz.de

C. Bischof (eds.), *Competence in High Performance Computing 2010*,
DOI 10.1007/978-3-642-24025-6_5,

the possible approaches to face these challenges in system monitoring of high performance computers. The main ideas involve a distributed software architecture with low communication overhead and reduced amounts of data stored by using data summarization techniques. These concepts have been developed into a prototype in test phase. The monitoring and analysis system prototype PerSyst Monitoring which we present in this paper has been developed at the Leibniz Supercomputing Center and ported on the Altix 4700 based on Itanium 2 architecture. It consists, for the time being, on hardware counter monitoring and analysis, taking to our advantage the rich facility given by the Itanium 2 architecture to provide detailed information on micro-architectural events. The supercomputer has a total of 9,728 processing cores which will be referred to as CPUs in the rest of this paper. Partitions are a group of CPUs with shared memory and in this machine there are 512 CPUs per partition. The Altix 4700, also called HLRB2, has a total of 19 partitions.

An important task of hardware counter analysis is to detect weak performance in running jobs. This is how we can distinguish which applications are running in the system and relate it to its performance properties. The largest part of the supercomputer is used for jobs which are sent via PBS batch system. Jobs go into the queuing system and are distributed according to certain priorities to run among partitions. They are identifiable by a unique job number, the JobID. There are also some special cases where cores are used outside of the batch system. These cases don't have an identification but are nonetheless monitored. PerSyst Monitoring calculates aggregated information for jobs, making it easier to identify jobs that require tuning.

This paper is organized in seven sections. The next section introduces the related work on monitoring systems. Section 3 addresses the software architecture details of PerSyst Monitoring. Section 4 presents the performance analysis and computations used by the tool. The information reduction of the obtained monitored data is explained in Sect. 5. Preliminary results are reported on the next section prior to conclusions and future work.

2 Related Work

The name, PerSyst Monitoring is derived from Periscope System Monitoring. Many of the features that have been taken for the development of the prototype, are from the profiling based tool Periscope [4]. Periscope is a scalable tool for analysing application performance. It enables a distributed online search for performance properties based on hardware counters as well as MPI properties. Periscope is not a monitoring system but some of the functionality of property processing in PerSyst Monitoring is similar to Periscope. A clear difference between them is that the latter one utilizes an instrumented code. It provides the possibility of defining a user region within the code [2]. There are also potential look-up regions within the instrumented section of the code which are automatically detected. These regions, also called phase regions, are analysed and refined when bottlenecks are detected in order to delimit the section of the code which presents a problem. In a high performance environment there is no inherent instrumentation of the applications running, the users shouldn't be requested to instrument their codes as this causes overhead. The analysis and monitoring is done without localising the code regions of potential stalls and weaknesses at each application. PerSyst Monitoring provides analysis at intervals at a system level, collectible per job, in order to have a first outlook on which applications need tuning. The closer examination at an application level would be the next step to our monitoring and analysis system by means of instrumentation and the use of a performance tool like Periscope.

Other system monitoring solutions have been developed. The one most related with our approach made by Nataraj et al. [7] for application monitoring has also a similar back end system monitoring by means of Supermon. This component, Supermon, has a scalability of up to 2048 [10] and handles client requests to extract performance information from the kernel. While PerSyst Monitoring is based on a cyclic system-wide measurement rationale, Supermon is driven by requests for localized data.

3 Agent Architecture and Communication

PerSyst Monitoring has been developed as a distributed software with a simple agent hierarchy: A High Level Agent, henceforth referred to as HLAgent, at the top of the hierarchy tree. The HLAgent communicates to Analysis Agents, from now on referred to as AAgents. The agent hierarchy is shown in Fig. 1. The socket based communication is implemented using the ACE (Adaptive Communication Environment) framework [8]. ACE is an object oriented framework designed to

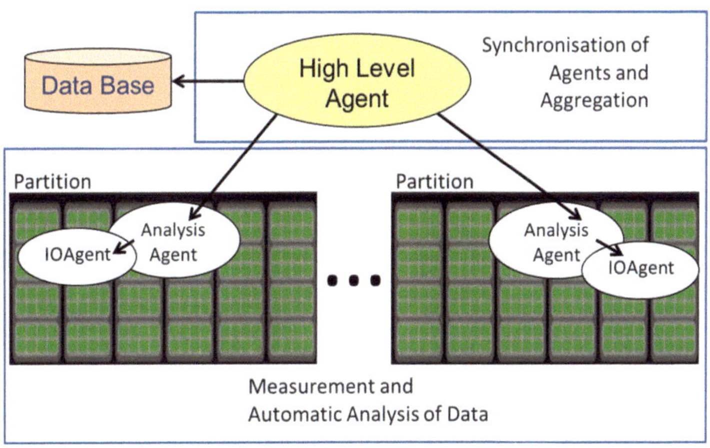

Fig. 1 Agent hierarchy

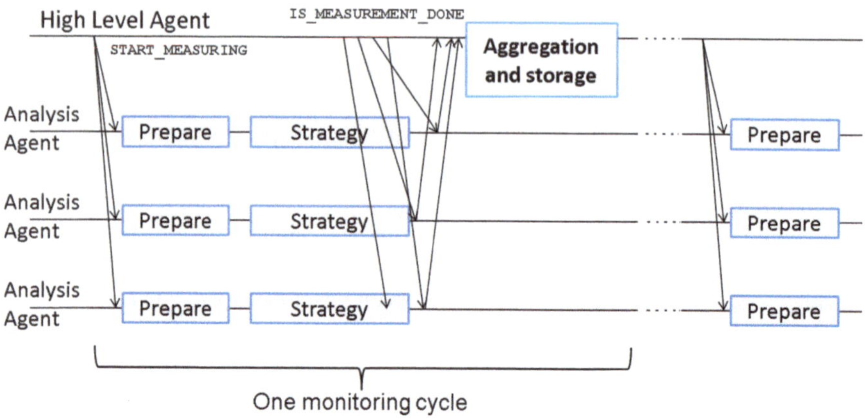

Fig. 2 Agent communication scheme

simplify the programming of interprocess communication, threading and memory management. It also serves as an API to general operating system services. All these qualities make it highly portable [9].

All AAgents are released by the HLAgent when it begins execution. They listen at a given port for instructions from the HLAgent before proceeding for measurements. The hosts to where each AAgent is to be distributed can be configured in the HLAgent. The HLAgent itself does not perform any measurements and is released in a login partition which is not monitored. An AAgent is intended to be in charge of monitoring only one partition of the supercomputer. It can monitor groups of CPUs with different strategies but there shouldn't be more than one AAgent dedicated at monitoring a partition. Within a partition the AAgent has the capability of analysing mutually exclusive groups of CPUs in parallel. This internal parallelisation was done in OpenMP. Every OMP thread will enter a parallel region and call a Performance Script. This script makes calls to *pfmon* [5]. The *pfmon* tool collects architecture specific hardware counter samples for all the existing Itanium processors.

A simple communication scheme as shown in Fig. 2 is used to send small packets of data between the HLAgent and the AAgents. This is done to synchronise the Agents to carry out measurements at the same time. The main HLAgent commands are the following:

- `START_MEASURING` command which is broadcasted to the AAgents to start measuring.
- `IS_MEASUREMENT_DONE`. The HLAgent sends a request to the AAgent to check if they are done with the measurements, each AAgent sends the status whether the measurement was carried out successfully or not.

Figure 2 shows one complete cycle of the communication and process scheme. The HLAgent sends the `START_MEASURING` command and the agents receive and process. Before executing the measurements the AAgent has a preparation step.

After a certain lapse of time the HLAgent sends the request `IS_MEASUREMENT_DONE` and waits with a timeout for the answers of the AAgents. The AAgents, in the ideal scenario, finish the measurement process, send a confirmation that the measurement is done and the HLAgent proceeds to execute the aggregation of information per JobID. Finally the storage is done by the HLAgent and the system is ready for a new monitoring cycle. The time between measurements is configurable and we estimated, given the nature of the jobs submitted on HLRB2, that measurements every 10 minutes are enough to detect inefficiencies in applications.

In real life we can expect all kinds of problems across the time span of a monitoring system. The communication protocol is used to overcome some of these problems. Failure in CPUs which may badly hamper the monitoring tasks of other healthy running devices can be expected. Not severe but a far more frequently emerging problem is the inability to measure some CPUs via *pfmon*.

For critical disruptions of any of the PerSyst Monitoring components, the HLAgent is responsible for the fault tolerance of the rest of the monitoring system. The HLAgent generates again the components which aren't responding anymore. For this purpose, there are implicit heartbeats with every communication attempt between the HLAgent and the AAgents. There are cyclical timeouts which are set to all the components. At the expiration of a timeout, if a heartbeat was not received, the HLAgent produces again the missing AAgent. The AAgent will also terminate on its own if it doesn't receive any command from the HLAgent after the timeout. So once the HLAgent terminates, the AAgents will exit and terminate soon after no command is received. That centralizes the matters of life and death of PerSyst Monitoring to the HLAgent. This component can then be configured to be monitored by other tools that probe for continuity of services.

For non-critical exceptions we also use the idea of implementing timeouts to continue the process of monitoring and for the time being we ignore those CPUs which do not deliver a result due to an exception within *pfmon*.

4 Performance Analysis

A property represents a deficiency or bottleneck, such as an unnecessary idleness of a micro-architectural hardware component. A deficiency can be determined by calculations made to a set of hardware counters, giving as a result the so called property value. If the property value is higher than a desired threshold then it is considered a severity. As an example consider the pipeline stalls. The hardware counters indicating a stall in the pipeline can point out very general causes or very specific ones. For instance, some of the counters that the Itanium 2 uses for counting pipeline stalls are: [1, 6]

- `BE_FLUSH_BUBBLE_ALL` for branch misprediction or exception.
- `BE_L1D_FPU_BUBBLE_ALL` for slow L1D hits,
- `BE_EXE_BUBBLE_ALL` for integer register dependencies or load instructions,

- `BE_RSE_BUBBLE_ALL` for register stack engine and
- `BACK_END_BUBBLE_FE` for instruction fetches.

Every single cause, whether it represents a very general one or a finer-level cause, is computed to produce a property value. For instance, the branch misprediction counter divided by all the cycles of a measured time is the property value for the branch-misprediction property. We define a threshold for every property, which is nothing else but a criteria to consider the property as inefficient or not. For the prototype the threshold holds constant in every property. The severity is the normalised difference between property value and threshold.

We take the idea of Periscope to differentiate the properties in a tree-like structure [3]. This is what we call a strategy. A strategy has a determined number of properties which in turn have a determined number of hardware counters involved. Before the measurement is done the AAgent collects all the counters belonging to the strategy in a set and sends it as input to the Performance Script. All the set of hardware counters involved in a strategy are measured at once or in more measurements via *pfmon* calls. Multiple *pfmon* calls are done due to a limitation of the architectural events that can be measured. There is a maximum number of counters which can be measured and some of them can't be measured together. To avoid these measurement conflicts the Performance Script calculates if a sequential measure is needed and if this is the case how the sets of counters should be grouped and used with *pfmon*. The AAgent receives the result from the Performance Script and it traverses the strategy tree for each CPU with its corresponding hardware counters to calculate the properties.

The HLAgent is the only agent with the objective of synchronising the AAgents, summarising results between partitions and ensuring continuity or persistence of the AAgents. The AAgents carry out local measurements on CPU groups at a given partition and process the measurements to follow a strategy.

5 Information Reduction

Information reduction is an important challenge for our monitoring prototype. We measure a complete strategy for all the CPUs in the supercomputer. From this preliminary information we calculate the property values and severities. The property value is a function of a set of hardware counters. Thus, the first information reduction is not to store the raw hardware counters but the property values. A second important filtering of information is to leave out unnecessary properties. We don't need to store a property value whose severity is not indicating an inefficiency problem. The strategy tree can have several levels of nodes. The absence of an inefficiency problem at a node makes the entire subtree of the corresponding node not worthy of further analysis or storage. This also applies to the root node, in case no inefficiency is found the entire strategy tree will not be stored.

The selected and condensed data is written to files. There is one file per CPU group and they are written in a shared file system across the partitions. The HLAgent is in charge of reading and processing the files to produce aggregated results over JobID with respect to the property value. This task is easily done by the HLAgent given that jobs mainly run on more than one partition, making it impossible for an AAgent to aggregate over JobID for jobs that are distributed over partitions. AAgents would need communication capabilities between themselves in order to aggregate over JobID.

Only the aggregated data is stored. The aggregated data consists of meaningful statistics per timestamp in a job, namely the average, the minimum (0-quantile) and at most 20 quantiles including 50-quantile and 100-quantile.

6 Results

A complete strategy was measured for a partition in HLRB2, the Altix 4700 supercomputer. This strategy, the stall cycle analysis, has three levels as shown in Fig. 3. The preliminary results of the strategy measured in timestamp 2010-02-09 6:30 in one partition are shown in Fig. 4. The CPUs 0 to 15 for the property `StallCyclesProp` have their severities equal to zero. Therefore, the following levels II and III will not be further analysed or stored for these CPUs. However, CPUs 16 to 23 in Fig. 4 presented a severity greater than zero. The strategy tree was further analysed to level II on these CPUs, given that `StallCyclesProp` leads to all the properties in level II. Further inefficiency problems in level II will lead to investigation of properties `BeL1dFpuBubbleL1dProp`, `BeL1dFpuBubble-FPUProp`, `BeExeBubbleGrgrProp`, `BeExeBubbleGrallProp` and `Be-ExeBubbleFrallProp` in level III.

Figure 5 shows the results obtained at timestamp 2010-02-09 6:40 for JobID 395036. This job uses 480 CPUs in total. This is the aggregation step per job which is done by the HLAgent after measurement is completed by the AAgents. We do

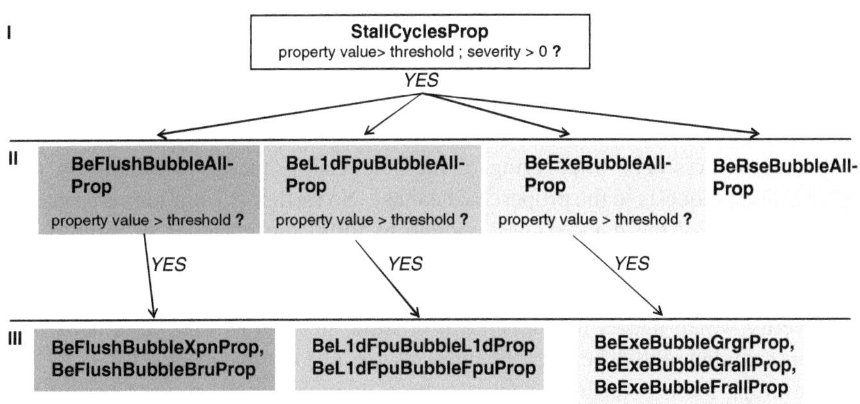

Fig. 3 Stall cycles strategy tree

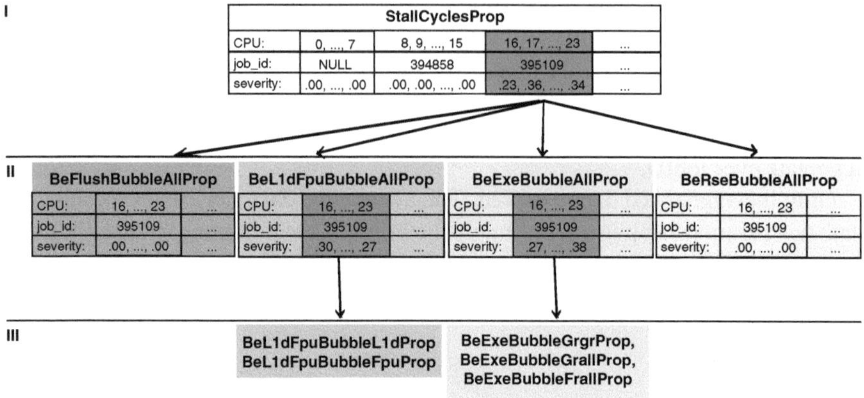

Fig. 4 Results for stall cycles strategy tree. Using timestamp 2010-02-09 6:30

not show the entire set of graphs of all the properties. The 480 CPUs are examined in the first level of the strategy and only those which are considered a bottleneck are examined on the second level of the strategy, i.e. 478 CPUs. From the second level we found property values which require further examination of properties `BeFlushBubbleAllProp` and `BeFlushBubbleBuProp` for 16 cores, which are not shown to simplify the layout of Fig. 5. Similarly, the refinement level goes into examining properties `BeExeBubbleGrallProp`, `BeExeBubbleFrallProp` and `BeExeBubbleGrgrProp` for 478 CPUs which are not shown. The graphs in Fig. 5 illustrate the percentage of CPUs (horizontal axis) which have a property value smaller or equal to the corresponding value shown in the vertical axis. The threshold line for calculating each severity is shown at each graph. The information reduction for this aggregation procedure was of 89%.

7 Conclusion and Future Work

The most outstanding advantage of the use of performance properties is that this leads to a better and faster interpretation of results than just reading the raw hardware counters. The uncovering of inefficiencies is readily available by just making direct requests to the properties database. No further calculations are needed and the quality of the data is good enough to fulfill the purposes of monitoring the supercomputer. If we need to reduce or increase the time to measure we can still compare the properties between samples. This is due to the normalisation of the severity which makes it independent of time. In contrast, the same hardware counters sampled with different time intervals can't be compared directly.

In average there is an important reduction of information as described in the results section. In the worst case scenario we need to store the same amount of information as we would if we store all the raw hardware counters, with the

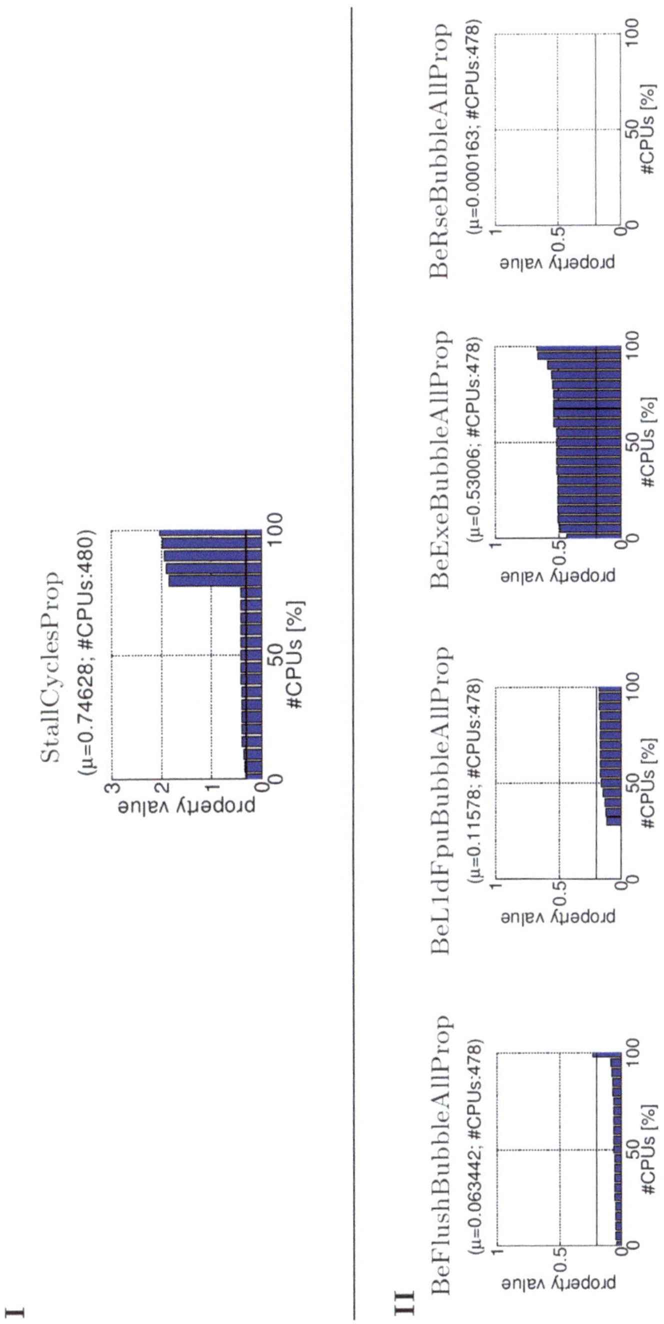

Fig. 5 Aggregation of JobID 395036 at timestamp 2010-02-09 6:40. μ is the arithmetic mean of the property values. #CPUs is the number of CPUs

difference, to our advantage, that the properties are easier to interpret. The best case scenario is no storage required, i.e. the properties did not show inefficiencies. The aggregation step provides a further reduction to our data. Both reduction strategies together produce, on average, about a 90% of reduction of the data.

Future work will include the porting and testing of PerSyst Monitoring to a petascale system. To achieve this, we plan to develop an interface that allows easy and fast development of other strategies for different processor architectures. The HLAgent could run into performance issues as it is a centralized unit that processes the condensed information of all the AAgents. An eventual need to parallelize its task is part of our outlook. In principle, the centralized work of storing in the data base could be undertaken by the AAgents themselves to avoid a performance decline.

Acknowledgements This work is funded by BMBF under the ISAR project, grant 01IH08005.

References

1. Gerndt, M., Fuerlinger, K.: Automatic performance analysis with periscope. Journal: Concurrency and Computation: Practice and Experience.Wiley InterScience. John Wiley & Sons, Ltd. (2009)
2. Gerndt, M., Fuerlinger, K., Kereku, E.: Periscope: Advanced techniques for performance analysis, parallel computing: Current & future issues of high-end computing. In: International Conference ParCo 2005, vol. 33 (2006). NIC Series ISBN 3-00-017352-8
3. Gerndt, M., Kereku, E.: Search strategies for automatic performance analysis tools. In: Euro-Par 2007, vol. LNCS 4641, pp. 129–138 (2007)
4. Gerndt, M., Strohhaecker, S.: Distribution of analysis agents in periscope on altix 4700. In: Proceedings of ParCo (2007)
5. HP: pfmon tool. www.hpl.hp.com/research/linux/perfmon/pfmon.php4
6. Intel: Introduction to Microarchitectural Optimization for Itanium 2 Processors (2002). URL http://cache-www.intel.com/cd/00/00/21/93/219348_software_optimization.pdf
7. Nataraj, A., Sottile, M., Morris, A., Malony, A., Shende, S.: Tauoversupermon: Low-overhead online parallel performance monitoring. In: Proceedings Euro-Par 2007, vol. LNCS 4641, pp. 85–96 (2007)
8. Schmidt, D.C.: The adaptive communication environment: Object-oriented network programming components for developing client/server applications. In: Proceedings of the 12th Annual Sun Users Group Conference, pp. 214–225 (1994)
9. Schmidt, D.C., Huston, D., Buschmann, F.: C++ Network Programming Vol. 1: Mastering Complexity with ACE and Patterns. Pearson Education (2002)
10. Sottile, M.J., Minnich, R.G.: Supermon: A high-speed cluster monitoring system. In: Proceedings of the IEEE International Conference on Cluster Computing, CLUSTER '02, pp. 39–. IEEE Computer Society, Washington, DC, USA (2002). URL http://portal.acm.org/citation.cfm?id=792762.793324

Design and Implementation of Distributed Meshes in HiFlow^3

Staffan Ronnås, Thomas Gengenbach, Eva Ketelaer, and Vincent Heuveline

Abstract This paper describes the conception and implementation of the mesh module created as part of the finite element software package HiFlow^3 [2]. With the aim of providing a general representation that can be reused in a wide range of applications, the code provides functions to work with meshes with different cell types in different dimensions through a uniform abstract interface. The decisions made during the design process were motivated by requirements coming from several applications. These requirements include the ability to handle large distributed meshes for cluster computing as well as local mesh refinement and coarsening for adaptive methods. A balance has been found between the need of high performance on the one hand and simple, maintainable code on the other.

1 Introduction

HiFlow^3 is a general-purpose finite element library, whose goal it is to greatly simplify the development of performant FEM simulations in a wide range of application areas, targeting different parallel hardware platforms including clusters and graphics processing units (GPUs). A central part of any such library is the representation of the computational mesh, which is used to approximate and discretize the geometric domain of the underlying mathematical model.

In contrast to what is common in FEM software specifically developed for a particular mathematical model, the mesh module in HiFlow^3 is generic in the sense that it can handle meshes in different dimensions and with different types of cells, including mixed-type meshes. The mesh module provides functionality for input and output from files, iteration over the entities in the mesh, refinement and coarsening

S. Ronnås · T. Gengenbach · E. Ketelaer · V. Heuveline (✉)
Engineering Mathematics and Computing Lab (EMCL), Karlsruhe Institute of Technology (KIT), Karlsruhe, Germany
e-mail: Staffan.Ronnas@kit.edu; Thomas.Gengenbach@kit.edu; Eva.Ketelaer@kit.edu; Vincent.Heuveline@kit.edu

C. Bischof (eds.), *Competence in High Performance Computing 2010*,
DOI 10.1007/978-3-642-24025-6_6, © Springer-Verlag Berlin Heidelberg 2012

of cells, and communication between meshes existing on different processors; none of which is specific to the cell types in the mesh or its dimension. This generality simplifies the development of FEM solvers that function with different types of mesh discretizations.

The paper first describes some basic mathematical ideas concerning the geometry and topology of meshes. This is followed by an explanation of the abstract interfaces that the module provides for interacting with meshes and mesh entities, as well as the first implementation of these interfaces, which is provided in the module. The next sections deal with three aspects of the mesh module in more detail: the generic description of cell types, the support for refinement and coarsening, and the management of parallel meshes on distributed-memory platforms.

2 Motivation

The vision of the HiFlow^3 project is to develop a versatile tool that offers the performance and flexibility required to solve large-scale applications efficiently and accurately. A special emphasis of the library is to tackle problems with high relevance in society in the research areas of energy, meteorology and environment, and medical engineering. The design of the mesh module, which is one of the cornerstones of the HiFlow^3 package, takes these aims into account and provides a trade-off between performance and flexibility that fits well into the overall concept of HiFlow^3.

Several different use cases drive the development of the mesh module. On the one hand, there is a need to handle the complex geometries which occur for instance in the *United Airways* project [7, 8, 13]. The aim of this project is the simulation of the full human respiratory system based on geometries, generated from CT-data. In this case, the representation of the geometry can be too large to fit into the memory of one machine. Therefore the mesh module uses a parallel data structure, and supports parallel input and output.

On the other hand, the use of adaptive finite element methods requires more freedom in the choice of cell types and the connectivities between the cells, in order to handle local refinement and coarsening of the mesh. The project *Goal Oriented Adaptivity for Tropical Cyclones* [9] is one example, where adaptive methods are used in order to solve problems in which phenomena on multiple scales interact with each other. Support for handling mesh adaptation in a user-friendly way is therefore an important component of the mesh module. For maximum flexibility, both conforming and non-conforming meshes can be used.

3 Geometry and Topology

In the context of HiFlow^3, we consider a mesh as being a partitioning of a domain into cells of a limited number of predefined shapes. The list of possible cell types is extensible, and there are already implementations for triangles and quadrilaterals in 2d, and tetrahedrons and hexahedrons in 3d. These shapes and the vertices are referred to as entities in the following. The cells are non-overlapping in the sense that their mutual intersections are lower-dimensional shapes, but the mesh does not need to be conforming, which means that the intersection of two neighboring cells does not have to be a sub-entity of both cells.

The geometry of the mesh is considered separately from its topology, in order to simplify the development of simulations involving for instance moving meshes. Currently, the geometry representation in the mesh module is simply an assignment of some coordinates to each vertex. This is the natural choice, as the vertex coordinates is the basic information provided by all mesh generators.

The topology of the mesh is described via the incidence relations (also called connectivities by some authors) between its entities. For the computation and representation of incidence relations, we closely follow the approach described in [14]. As in that paper, we denote the incidence relation between entities of dimension d_1 and those of dimension d_2 in a mesh by $d_1 \rightarrow d_2$. What this incidence relation means depends on whether d_1 is larger than, smaller than, or equal to, d_2.

- $d_1 > d_2$: the d_2-entities contained in each d_1-entity (e.g. $3 \rightarrow 1$ the edges of a 3d-cell).
- $d_1 < d_2$: the d_2-entities containing each d_1-entity (e.g. $1 \rightarrow 3$ the cells sharing an edge).
- $d_1 = d_2$; $d_1, d_2 > 0$: the d_2-entities sharing at least one vertex with each d_1-entity.

Starting from the input connectivity $D \rightarrow 0$, where D is the topological dimension of the cells of the mesh, all connectivities can be computed through the combination of three basic algorithms:

- `build`: computes $d \rightarrow 0$ and $D \rightarrow d$ from $D \rightarrow D$ and $D \rightarrow 0$, for $0 < d < D$.
- `transpose`: computes $d_1 \rightarrow d_2$ from $d_2 \rightarrow d_1$, for $d_1 < d_2$.
- `intersect`: computes $d_1 \rightarrow d_2$ from $d_1 \rightarrow d_3$ and $d_3 \rightarrow d_2$, for $d_1 \geq d_2$.

These algorithms, together with the algorithm that combines them to compute a given connectivity is described in [14]. The special case $0 \rightarrow 0$ has been omitted, since it has not been implemented in HiFlow^3.

4 Abstract Interface Classes

The use of abstract interfaces is the cornerstone of modular programming, which enables a decoupling between different parts of a program, allowing each to vary separately. The mesh module provides an abstract interface to its services via a collection of classes with separate responsibilities. The relationships between these classes is shown in Fig. 1.

The `Mesh` class represents a computational mesh. It is an abstract base class, and hence cannot be instantiated. Instead, different concrete mesh implementations can

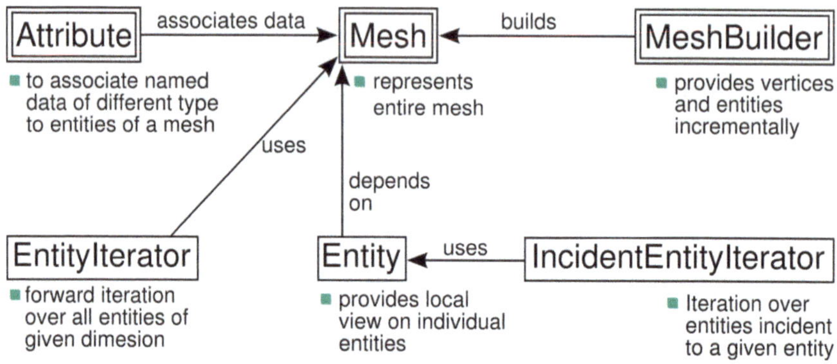

Fig. 1 Mesh interface classes and their relations

be created, possibly with different performance characteristics, and different levels of generality. The use of dynamic polymorphism instead of static polymorphism based on templates, which is the current trend in scientific C++ programming, is motivated through the higher degree of flexibility (the implementation can be chosen at runtime), and the simpler code that results. The price to pay is the overhead of virtual function calls, which the authors believe to be small for typical use cases. As a comparison, an example of the use of static polymorphism for mesh handling can be found in the dune-grid module of the DUNE project [3, 4].

An important characteristic of the `Mesh` class is that its public interface contains almost exclusively `const` functions, meaning that a `Mesh` can be considered to be an immutable object. This design choice was made to make it easier to reason about the validity of the state of the `Mesh` objects. The only time at which a `Mesh` can be modified is when it is constructed. This responsibility is given to implementations of the `MeshBuilder` interface, which in accordance to the Builder design pattern [6], lets the user build the mesh by providing its vertices and entities incrementally, before obtaining the finished `Mesh` object via the `build()` method.

An exception to the immutability of the `Mesh` object is the attributes mechanism. Similar to the `MeshFunction` class described in [14], the `Attribute` class provides a way for the user to associate named data of different types to the entities of the `Mesh`. This concept is also used to store results associated with different mesh operations, such as refinement.

The `Mesh` abstract base class represents an entire mesh. In many algorithms, however, it is useful to work with a local view of individual entities. This is provided through the `Entity` class, which uses the `Mesh` interface to obtain the data associated with a single entity of the mesh. Even though it is a concrete class and can be instantiated, it only depends on the `Mesh` interface, and can thus be used with all possible `Mesh` implementations.

Random access to the entities of a `Mesh` is realized through the function `Mesh::get_entity()`, but it is also possible to iterate over entities in two ways. Forward iteration over all entities of a given dimension in the `Mesh` is implemented through the class `EntityIterator` and the functions `Mesh::begin()` and `Mesh::end()`. Iteration over the entities incident to a given entity is provided through the class `IncidentEntityIterator`, which is obtained from the pair of functions `Entity::begin_incident()` / `Entity::end_incident()`. Like the `Entity` class, the iterator classes are independent of the concrete `Mesh` implementation used. More complex types of iteration can be implemented with the help of the Boost.Iterator framework [1], which these iterator classes build upon.

5 Mesh Implementation

There is at the moment one implementation of the interface defined by the `Mesh` class, which is provided by the `MeshDbView` and `RefinedMeshDbView` subclasses. The second class derives from the first, and represents a mesh that is a refinement of another mesh. In addition to the data stored in `MeshDbView`, `RefinedMeshDbView` also has the refinement history associated to it, and overloads some of the functions to use this refinement history.

Most of the functionality is however provided by the `MeshDbView` class, which heavily depends on a third class, `MeshDatabase`. The `MeshDatabase` class manages a unique numbering for all entities over a set of meshes. A single `MeshDatabase` object is shared between all meshes in a refinement hierarchy, and between a mesh and its boundary mesh. In accordance with its name, a `MeshDbView` object represents a limited view of the `MeshDatabase`: either a level in the refinement hierarchy, or the boundary of a mesh.

The `MeshDatabase` class manages the entity-vertex and vertex-entity connectivities $d \to 0$ and $0 \to d$ for all existing entities. On demand, the `MeshDbView` class can compute and cache the restriction of these connectivities to the entities belonging to a particular mesh. It can also compute the other connectivities upon request.

The `MeshDatabase` class provides set semantics for its entities, meaning that if an entity is added several times, it only exists once and always receives the same id. This is implemented with the help of an additional structure, the `VertexSearchTable`, which makes it possible to search for a vertex by its coordinates. The underlying data structure is simply an array of the vertex ids sorted by the distance of the vertex to the origin. The search for a given vertex consists of first finding all vertices, which are at the same distance from the origin (within a small tolerance), and then performing a linear search over these vertices. Although the worst-case efficiency of this structure is not optimal, it has proven to be quite fast when used together with typical simulation data.

The `VertexSearchTable` makes it easy to enforce set semantics for vertices. When a vertex is added to the `MeshDatabase`, a search is performed first to see whether it already exists. If so, the id of the existing vertex is returned instead of creating a new vertex. For the entities of dimension larger than 0, the corresponding search is performed using the $0 \to d$ connectivities, which, unlike the other connectivities, are sorted. Looking up an entity, specified by its vertex ids, consists in computing the intersection of the sets of entities connected to each of its vertices. If the entity does not exist, this intersection will be empty, and otherwise it will be the set containing the id of the sought entity.

Building new `MeshDbView` objects is done through the `MeshDbViewBuilder` class, which implements the interface defined by the `MeshBuilder` class. The `MeshDbViewBuilder` holds a reference to the `MeshDatabase` to which new entities should be added. The fact that the `MeshDatabase` provides set semantics for adding vertices and entities, makes the implementation of the `MeshDbViewBuilder` extremely simple: the same vertex or entity can be added several times, but will always receive the same id, which saves the `MeshDbViewBuilder` class from having to keep track of which entities have already been added.

6 Cell Types

The different types of cells (lines, triangles, quadrilaterals, tetrahedrons and hexahedrons) are described through the `CellType` class hierarchy. The base class, `CellType`, stores information about the local $D \to d$ connectivities and the possible refinements of a cell type. By combining these two pieces of information, the connectivities of the refined sub-cells are also derived, which is the basis for the handling of non-conforming meshes. Each subclass of `CellType` simply implements some callback functions that return the specific information for that cell type, and is only used for initialization. The data is stored in the base class.

A refinement is specified in three steps. Firstly, so-called "refined vertices" are added to the cell type. A refined vertex is defined as the barycenter of a set of existing vertices. In a second step, one can then define the "sub-cells" by specifying the set of vertices (regular or refined) that it contains. Finally a refinement is defined as the set of sub-cells that should be produced by the refinement. This description allows to refine to a different cell type, e.g. quadrilaterals to triangles.

The connectivities for the sub-cells and its sub-entities are computed automatically using the `CellType` definitions of the sub-cells. The central idea is that "refined" entities are numbered consecutively, starting after the corresponding "regular" entities. In each cell type, cell 0 is the cell itself, and cells $(1, \dots)$ are the sub-cells that can be included in a refinement. Similarly, if vertices $(0, \dots, \mathrm{N})$ are the vertices of cell 0 (i.e. the regular vertices), one can add refined vertices $(\mathrm{N}+1, \dots)$, which can be used to define the sub-cells.

In this way, it is possible to compute the local connectivities between sub-cells, and extract all information necessary both for refinement and for dealing with non-conforming meshes. Numbering the entities consecutively makes it easier to deal with entities and sub-entities in a uniform way.

7 Refinement and Coarsening

Refinement and coarsening of a mesh is performed with the `Mesh::refine()` function. It takes as input an array indicating for each cell if it should be coarsened, refined (and if so, how), or left as it is. The function builds the refined mesh, and returns it to the caller. Figures 2 and 3 show the refinement of a cube and a tetrahedron, respectively.

In the `MeshDbView` implementation, `RefinedMeshDbView`, a specialized subclass, is used to represent refined meshes. This class keeps track of the refinement history by storing an array of pointers to all ancestor meshes, as well as two attributes that indicate for each cell in which ancestor mesh the parent cell lives, and its index, local to that mesh. This makes it possible to access the parent of a cell, even if one only has access to the refined mesh.

In addition to these two attributes, we have also found it useful to store the "sub-cell number", i.e. the index of the sub-cell in the parent's cell type. This information is used both for boundary extraction and handling of non-conforming meshes.

During refinement, some cells can also be coarsened. This means that they are replaced by their parents. There is some ambiguity about what to do when one cell is marked to be coarsened, while another cell sharing the same parent (a sibling) is not marked to be coarsened. One could for instance either force a coarsening of all sibling cells, or require all sibling cells to be marked for coarsening, before it can take place. We follow the second path, by searching for "permitted ancestor" cells for all cells that are marked to be coarsened. A "permitted ancestor" is a cell

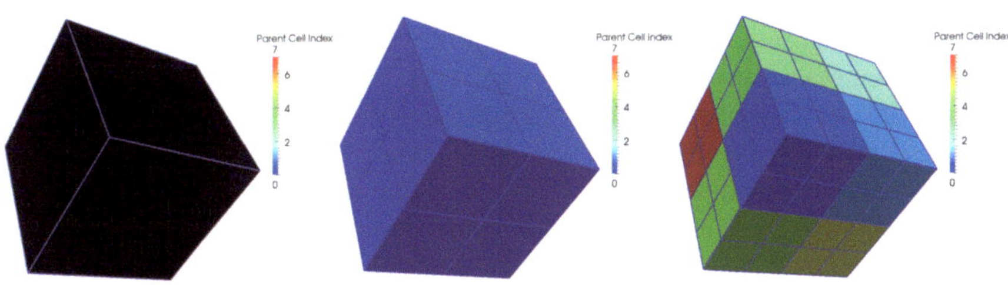

Fig. 2 Refinement of a cube with the parent cell index attribute.

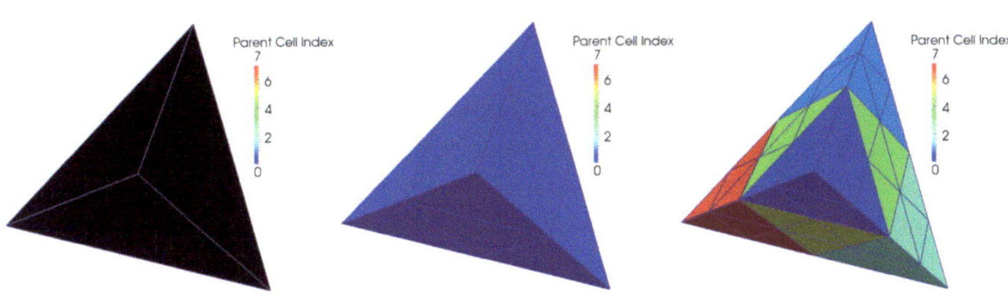

Fig. 3 Refinement of a tetrahedron with the parent cell index attribute.

in an ancestor mesh, such that all its children in the current mesh are marked to be coarsened. If no "permitted ancestor" is found, the coarsening mark on the cell is removed, and it is left untouched. If one or several "permitted ancestors" are found, all their children are coarsened. This means that coarsening of large areas can be carried out with one coarsening step.

8 Distributed Meshes

The mesh module offers the possibility to work with distributed meshes. In the current implementation, a distributed mesh is a set of `Mesh` objects, one on each process in the communicator group. The communication is handled by functionality external to the `Mesh` classes, which themselves are not aware that they are part of a larger, global mesh. The advantage of this is that all `Mesh` functions are local to a process and do not require communication, which facilitates their implementation and use. Code reuse is also possible, since the communication code works with the `Mesh` interface, and not the individual implementations.

In order to communicate between processes, it is necessary to know how the parts of the global mesh are connected. This is done via the `SharedVertexTable` class, which contains the information about which of the local vertices are shared with what other processes, and what the remote id is of each shared vertex. Having this information for vertices, it is possible to also identify shared entities of higher dimension.

If the mesh has been read in from a parallel VTK file [15], the information provided by the `SharedVertexTable` is not available. Thus it has to be computed in a global communication by the function `update_shared_vertex_table()`, which exchanges the coordinates of all vertices of a mesh with all other processes, looks up the ids of the received vertex coordinates locally via the `MeshDatabase`, and communicates this information back to all other processes. This is a potentially expensive operation, but necessary in the case that no other information is available.

The communication of mesh entities is performed via a simple serialization procedure. A set of entities defined by an `EntityIterator` range is packed into an `EntityPackage` object, which is then communicated. Two modes of communication, scatter and non-blocking point-to-point, have been implemented in the `MpiScatter` and `MpiNonBlockingPointToPoint` classes, respectively. On the receiving end, the `EntityPackage` object is rebuilt, and can then be used together with a `MeshBuilder` to reconstruct the mesh on the remote process. Again, for the `MeshDbView` implementation, the set semantics of the `MeshDatabase` greatly simplify this procedure.

The computation and communication of ghost cells have been built on top of this framework. A higher-level function takes a local mesh on each process, and creates a new mesh containing one level of ghost cells from the neighboring processes. All cells are marked with attributes that indicate the owning process and the index of the cell on that process. Figure 4 shows a distributed mesh describing the geometry of a human nose.

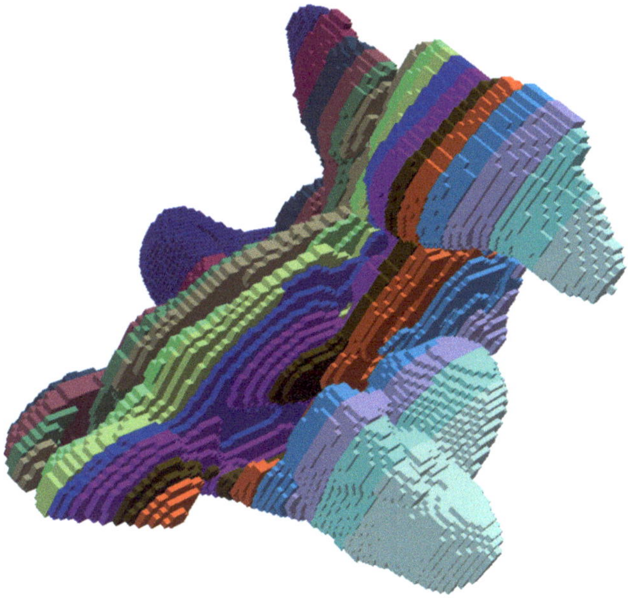

Fig. 4 A mesh of a human nose distributed in 16 stripes. One stripe and one layer of ghost cells exist on each process.

There is also support for computing the partitioning of a mesh, i.e. deciding how it is to be distributed over the processes. The `GraphPartitioner` abstract base class provides an interface, for which there are currently two implementations. One is provided through an interface to the well-known library METIS [12], and the second is a "naive" implementation, which partitions based on the numbering of the cells. At the moment, these partitioners work on the local mesh only, and in practice one reads in the mesh on one process, computes the partition, and then distributes the parts to the other processes.

9 Future Work

The first implementation of the mesh interface described in this paper has already been used for solving several different problems on varying types of meshes, both in 2d and 3d. It is, however, still a work in progress, and will continue to be improved, both with respect to functionality and performance.

We plan to implement a more complete description of the geometry, which can greatly improve the accuracy of solutions on domains with curved boundaries. An interesting development in this direction is the idea of *Isogeometric Analysis*, in which one aims to combine the techniques of computer-aided design (CAD) and finite element simulations [10]. Support for moving meshes, which requires mainly modifications to the geometry, and quality criteria for meshes are also under consideration for inclusion in the module.

In order to achieve scalability for parallel adaptive computations, we furthermore aim to implement parallel partitioning and redistribution of mesh entities. For this task, there are existing parallel graph partitioning libraries such as ParMETIS [11] and PT-SCOTCH [5], whose services could be leveraged.

There is room for performance improvements, especially in what concerns the modifying functionality (refinement/coarsening and communication of mesh entities). The approach involving the `MeshDatabase` class and the `MeshBuilder` interface is not optimal in this respect, and we consider replacing these constructions with more dedicated and efficient data structures for use with very demanding computations. This will, however, require a shift towards more complexity in the code.

10 Conclusion

We have described the design and implementation of the mesh module in HiFlow^3. The library can be used in a uniform way for meshes of different dimensions and cell types. Furthermore, it is suitable for parallel computation with distributed meshes, and it supports complex refinement and coarsening, as well as non-conforming meshes. An abstract interface provides easy access to, and manipulation of, mesh objects as a whole, as well as individual entities. The use of this interface is motivated through the code modularity that it provides, and the way in which it simplified the implementation of the mesh module itself.

References

1. Abrahams, D., Siek, J., Witt, T.: The Boost.Iterator Library (2003). http://www.boost.org/doc/libs/release/libs/iterator/
2. Anzt, H., Augustin, W., Baumann, M., Bockelmann, H., Gengenbach, T., Hahn, T., Heuveline, V., Ketelaer, E., Lukarski, D., Otzen, A., Ritterbusch, S., Rocker, B., Ronnas, S., Schick, M., Subramanian, C., Weiss, J.P., Wilhelm, F.: HiFlow^3 – A Flexible and Hardware-Aware Parallel Finite Element Package. EMCL Preprint Series (2010). URL http://www.emcl.kit.edu/preprints/emcl-preprint-2010-06.pdf
3. Bastian, P., Blatt, M., Dedner, A., Engwer, C., Klöfkorn, R., Ohlberger, M., Sander, O.: A Generic Grid Interface for Parallel and Adaptive Scientific Computing. Part I: Abstract Framework. Computing **82**(2-3), 103–119 (2008)
4. Bastian, P., Blatt, M., Dedner, A., Engwer, C., Klöfkorn, R., Ohlberger, M., Sander, O.: A Generic Grid Interface for Parallel and Adaptive Scientific Computing. Part II: Implementation and Tests in DUNE. Computing **82**(2-3), 121–138 (2008)
5. Chevalier, C., Pellegrini, F.: PT-Scotch: A tool for efficient parallel graph ordering. Parallel Computing **34**(6-8), 318 (2008)
6. Gamma, E., Helm, R., Johnson, R.E., Vlissides, J.: Design Patterns: Elements of Reusable Object-Oriented Software. Addison-Wesley, Reading, MA (1995)

7. Gengenbach, T., Henn, T., Heppt, W., Heuveline, V., Krause, M.J.: United Airways: Numerical Simulation of the Human Respiratory System (2010). http://www.united-airways.eu
8. Gengenbach, T., Heuveline, V., Krause, M.: Numerical Simulation of the Human Lung: A Two-scale Approach. In: BMT 2010 – Reguläre Beiträge (BMT 2010 Reguläre Beiträge). Rostock-Warnemünde, Germany (2010)
9. Goal Oriented Adaptivity for Tropical Cyclones (2010): http://www.emcl.kit.edu/goa_for_tcs.php
10. Hughes, T., Cottrell, J., Bazilevs, Y.: Isogeometric analysis: CAD, finite elements, NURBS, exact geometry and mesh refinement. Computer Methods in Applied Mechanics and Engineering **194**(39-41), 4135–4195 (2005)
11. Karypis, G., Kumar, V.: A parallel algorithm for multilevel graph partitioning and sparse matrix ordering. J. Parallel Distrib. Comput. **48**(1), 71–95 (1998)
12. Karypis, G., Kumar., V.: A Fast and Highly Quality Multilevel Scheme for Partitioning Irregular Graphs. SIAM Journal on Scientific Computing **20**(1), 359–392 (1999)
13. Krause, M.J., Gengenbach, T., Heuveline, V.: Hybrid Parallel Simulations of Fluid Flows in Complex Geometries: Application to the Human Lungs. In: Proceedings of Euro-Par 2010 Workshops. Revised Selected Papers Springer-Verlag, Vol. 6586, 209-216 (2010)
14. Logg, A.: Efficient Representation of Computational Meshes. International Journal of Computational Science and Engineering **4**(4), 283–295 (2009)
15. VTK – The Visualization Toolkit: http://www.vtk.org/

Static and Dynamic Properties of Curved Vapour-Liquid Interfaces by Massively Parallel Molecular Dynamics Simulation

M.T. Horsch, S.K. Miroshnichenko, J. Vrabec, C.W. Glass, C. Niethammer, M.F. Bernreuther, E.A. Müller, and G. Jackson

Abstract Curved fluid interfaces are investigated on the nanometre length scale by molecular dynamics simulation. Thereby, droplets surrounded by a metastable vapour phase are stabilized in the canonical ensemble. Analogous simulations are conducted for cylindrical menisci separating vapour and liquid phases under confinement in planar nanopores. Regarding the emergence of nanodroplets during nucleation, a non-equilibrium phenomenon, both the non-steady dynamics of condensation processes and stationary quantities related to supersaturated vapours are considered. Results for the truncated and shifted Lennard-Jones fluid and for mixtures of quadrupolar fluids confirm the applicability of the capillarity approximation and the classical nucleation theory.

M.T. Horsch
Molecular Systems Engineering (MSE), Centre for Process Systems Engineering, Imperial College London, South Kensington Campus, London SW7 2AZ, England

Thermodynamik und Energietechnik (ThEt), Institut für Verfahrenstechnik, Universität Paderborn, Warburger Str. 100, 33098 Paderborn, Germany

S.K. Miroshnichenko · J. Vrabec (✉)
Thermodynamik und Energietechnik (ThEt), Institut für Verfahrenstechnik, Universität Paderborn, Warburger Str. 100, 33098 Paderborn, Germany
e-mail: jadran.vrabec@uni-paderborn.de

C.W. Glass · C. Niethammer · M.F. Bernreuther
Höchstleistungsrechenzentrum Stuttgart (HLRS), Nobelstr. 19, 70569 Stuttgart, Germany

E.A. Müller · G. Jackson
Molecular Systems Engineering (MSE), Centre for Process Systems Engineering, Imperial College London, South Kensington Campus, London SW7 2AZ, England

C. Bischof (eds.), *Competence in High Performance Computing 2010*,
DOI 10.1007/978-3-642-24025-6_7,

1 Introduction

The influence of curvature on the properties of a nanodroplet, as opposed to an interface that is planar (on the molecular level), is hard to capture experimentally. Yet it is important for refrigeration and energy technology as well as meteorology to understand fluid interfaces with extremely high curvatures because they characterize the onset of condensation and boiling processes.

Beginning in the 1930s with the work of Verschaffelt [1], researchers became aware of the necessity of taking the internal structure of fluid interfaces into account. They increasingly looked beyond the picture of a discrete dividing surface as postulated by Gibbs [2]. In the subsequent years, this led to the theoretical work of Guggenheim [3] and Tolman [4] which, nonetheless, was mostly still based on effective radii and hence on discretization.

Today, molecular dynamics (MD) simulation provides a means of accessing the internal structure and the non-equilibrium behaviour of vapour-liquid interfaces directly, on the basis of physically sound but algebraically simple effective pair potentials. For the truncated and shifted Lennard-Jones (LJ·TS) potential [5]

$$u(r) = \begin{cases} 4\varepsilon\left[\sigma^{12}(r^{-12} - r_{\mathrm{c}}^{-12}) + \sigma^{6}(r_{\mathrm{c}}^{-6} - r^{-6})\right], & \text{for } r < r_{\mathrm{c}}, \\ 0, & \text{for } r \geq r_{\mathrm{c}}, \end{cases} \tag{1}$$

with a cutoff radius of $r_{\mathrm{c}} = 2.5\ \sigma$, which constitutes a reliable model for the noble gases and methane [6], fluid phase boundaries have been simulated by several groups in the recent past [6–10]. Molecular simulation is particularly suitable for investigating metastable states, cf. Fig. 1. Virial isotherms that accurately describe the conditions of fluid phase coexistence at planar or curved interfaces

$$\frac{p}{T} = \sum_{j=1}^{5} b_j \rho^j, \tag{2}$$

which were correlated to the present data, are given in Table 1. Therein, p is the pressure, T is the temperature, ρ is the density, and the convention $k = 1$ is used. On this basis, the present work regards both equilibrium and non-equilibrium phenomena for vapour-liquid interfaces of the LJ·TS fluid. Nucleation in supersaturated vapours is considered for mixtures of quadrupolar fluids as well. For a more detailed exposition, the reader is pointed to four recent articles [10–13].

2 Curved Fluid Interfaces in Equilibrium: Theory

The Tolman [4] approach to curved vapour-liquid interfaces is based on comparing different effective radii of a droplet, namely the Gibbs adsorption radius R_ρ, for which the interfacial excess density is zero, and the Laplace radius R_{L} from

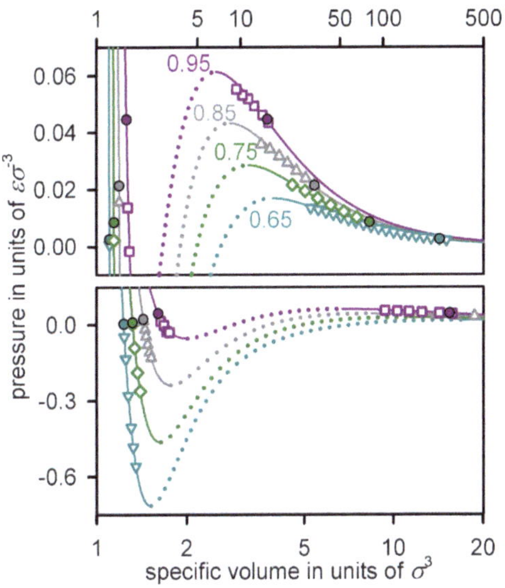

Fig. 1 Isotherms in a pressure-volume diagram for the LJ·TS fluid as determined by canonical MD simulation at temperatures of $T = 0.65$ (*down triangles*), 0.75 (*diamonds*), 0.85 (*upward triangles*), and 0.95 ε (*squares*) as well as the saturated states (*bullets*) according to Vrabec et al. [6] in comparison with the present fifth-order virial expansion (*continuous and dotted lines*), cf. Table 1 and Eq. (2). Results in the vicinity of the spinodal line are not shown here, and no such values were used for adjusting the virial coefficients, to ensure that only states unperturbed by nucleation were taken into account

Table 1 Virial coefficients for the LJ·TS fluid as determined from a fit to the MD simulation results shown in Fig. 1 and to the saturated vapour and liquid densities determined by Vrabec et al. [6]. The spinodal densities $(\rho')^\#$ and $(\rho'')^\#$ for liquid and vapour, respectively, were determined from the virial expansion, cf. Eq. (2)

T	$-b_2$	b_3	$-b_4$	b_5	ρ'	ρ''	$(\rho')^\#$	$(\rho'')^\#$
0.65	11.7675	44.5866	96.9625	71.4351	0.813	0.00406	0.660	0.0592
0.7	9.77572	34.176	76.4866	59.4954	0.787	0.00728	0.636	0.0740
0.75	8.43697	27.7315	62.373	50.3464	0.759	0.0124	0.613	0.0886
0.8	7.33394	21.854	41.1349	40.3329	0.730	0.0198	0.588	0.103
0.85	6.48592	18.3318	40.0252	34.6962	0.699	0.0304	0.564	0.119
0.9	5.44587	12.3036	25.0989	23.6305	0.664	0.0446	0.532	0.134
0.95	4.97043	10.0411	17.1387	16.0653	0.622	0.0648	0.499	0.149
1	4.67665	9.83155	15.6063	13.8778	0.571	0.0962	0.466	0.174

$$\gamma = \frac{1}{2} R_{\mathrm{L}}(p_\iota - p), \tag{3}$$

i.e. the Laplace equation in terms of the surface tension γ and the pressure p_ι inside a droplet containing ι molecules under equilibrium conditions. The deviation between these radii, the Tolman length

$$\delta = R_\rho - R_\mathsf{L}, \tag{4}$$

characterizes the curvature dependence of the surface tension [4]. The *capillarity approximation*, which postulates the droplets to be spherical and incompressible with a curvature independent surface tension, assumes δ to be zero.

According to the formalism employed by Buff [14] and Kondo [15], the surface tension becomes minimal if it is evaluated with respect to R_L. It can be shown that this assertion is only valid if the interfacial area F is proportional to R_L^2. However, both mechanical and thermodynamic equilibrium conditions for a droplet containing ι molecules imply

$$R_\mathsf{L} = 2\left(\frac{\partial V_\iota}{\partial F}\right)_{N,V,T}, \tag{5}$$

where V_ι and V are the volumes occupied by the droplet and the system as a whole, respectively. This only agrees with $F \sim R_\mathsf{L}^2$ if curvature effects cancel out.

For cylindrical interfaces, the surface tension varies with the radius according to

$$\left[\left(\frac{\partial \ln R_\mathsf{L}}{\partial \ln \gamma}\right)_T - 1\right]^{-1} = \frac{\delta}{R_\mathsf{L}} + \frac{\delta^2}{2R_\mathsf{L}^2}, \tag{6}$$

an expression that is similar and analogous to Tolman's equation for droplets, where R_L is defined to be positive for convex and negative for concave menisci. By combining the Young equation [16] with an expansion of γ to first order in $1/R_\mathsf{L}$, the contact angle ϑ of a fluid confined in a planar nanopore is obtained as

$$\cos\vartheta = \left(\frac{\gamma_\infty}{\Delta\gamma_\mathrm{s}} + \frac{\delta_\infty}{R_\mathrm{min}}\right)^{-1}. \tag{7}$$

Therein, γ_∞ is the surface tension of the planar vapour-liquid interface, δ_∞ is the Tolman length in the planar limit, and the pore diameter is $2R_\mathrm{min}$, while $\Delta\gamma_\mathrm{s}$ indicates the difference between the specific surface energies of the substrate when it is in contact with the vapour and the liquid, respectively.

3 Curved Fluid Interfaces in Equilibrium: MD Simulation

Using the *ls*1 *mardyn* MD program [19], equilibrium states involving droplets and cylindrical menisci were simulated for the LJ·TS fluid.

Vapour-droplet equilibrium MD simulations were conducted for droplets containing on the order of 100 to 1,000 molecules in the canonical ensemble, where such equilibria can be stable – as opposed e.g. to the grand canonical ensemble where this corresponds to a free energy maximum. The droplet size was evaluated according to a version of the cluster criterion of ten Wolde and Frenkel [20] with the connectivity radius $R_\ell = 1.5\,\sigma$ and coordination numbers $j \geq 4$ defining the

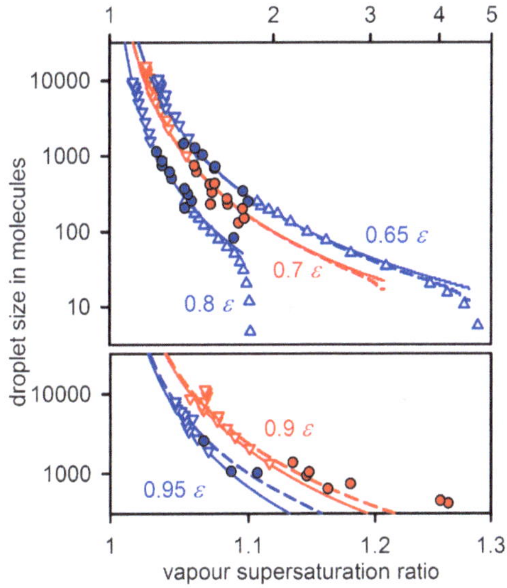

Fig. 2 Droplet size $\iota^\star$ over the supersaturation ratio S_μ (in terms of the chemical potential) for vapour-droplet equilibria of the LJ·TS fluid (*bullets*) from the present work, (*down triangles*) according to Vrabec et al. [6], (*upward triangles*) according to Napari et al. [17], and following the capillarity approximation (*continuous lines*) as well as the Laaksonen et al. [18] model (*dashed lines*)

liquid phase. The present results correspond to moderately supersaturated vapours, cf. Fig. 2, and are consistent with the results of Vrabec et al. [6] on larger droplets as well as the study of Napari et al. [17] covering vapours at pressures that approach the spinodal line. In the intermediate regime, the droplet size in equilibrium generally agrees well with the capillarity approximation.

Cylindrical interfaces were investigated by simulating liquid slabs, cf. Fig. 3, confined between two planar and layered walls represented by coupled harmonic oscillators. The equilibrium positions of the wall atoms were aligned according to a hexagonal structure with an interatomic distance of 0.3816 σ, corresponding to the bond length in graphite expressed in terms of the σ parameter value for methane. Both the fluid-fluid and the fluid-wall interactions were modelled by the LJ·TS potential, employing equal size parameters $\sigma_{\mathrm{fw}} = \sigma$ in both cases, while the dispersive energy between fluid molecules and wall atoms

$$\varepsilon_{\mathrm{fw}} = \zeta\varepsilon, \tag{8}$$

was systematically varied. The arithmetic mean of the saturated vapour and liquid densities was selected as a criterion for detecting the phase boundary. A circle was adjusted to the resulting profile at distances between 2 and 11 σ from the wall, cf. Fig. 4, and the tangent to this circle at a distance of 1 σ from the wall was examined to determine the contact angle. Qualitatively, the contact angles obtained by the present MD simulations are captured by Eq. (7), assuming a proportionality law for

$$\Delta\gamma_{\mathrm{s}} = \mathsf{K}_\gamma(\rho' - \rho'')(\zeta - \zeta_0), \tag{9}$$

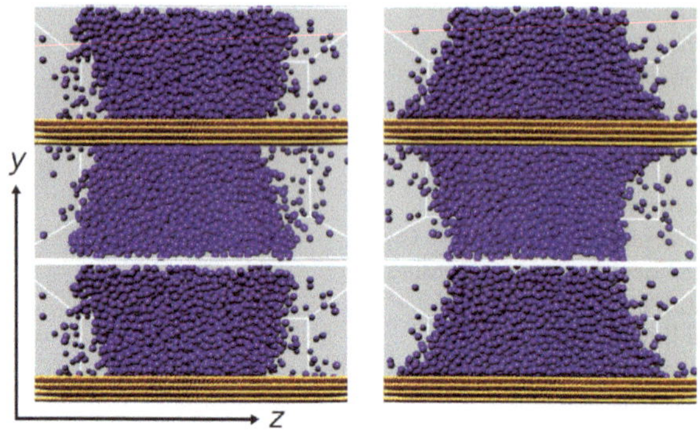

Fig. 3 Simulation snapshots for the reduced fluid-wall dispersive energy ζ of 0.09 (*left*) and 0.16 (*right*) at a temperature of 0.73 ε. The upper half is reproduced in the bottom to illustrate the effect of the periodic boundary condition

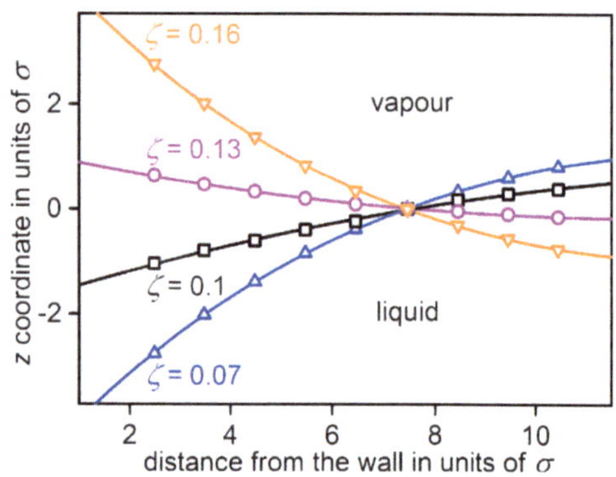

Fig. 4 Vapour-liquid interface profiles for the reduced fluid-wall dispersive energy ζ of 0.07 (*upward triangles*), 0.10 (*squares*), 0.13 (*circles*), and 0.16 (*down triangles*) at a temperature of 0.82 ε

cf. Fig. 5. The magnitude of the fluid-wall dispersion for which ϑ becomes rectangular ($\zeta_0 = 0.118$) was found to be temperature independent.

4 Homogeneous Vapour to Liquid Nucleation: Theory

The foundations of the classical nucleation theory (CNT), concerning the first step of a first-order phase transition in the bulk of a metastable phase, were laid by Volmer and Weber [21] as well as Farkas [22]. On the basis of the capillarity approximation, the free energy of formation ΔA of a droplet containing ι molecules in the thermodynamic limit (i.e. for an infinitely large vapour phase at a constant supersaturation ratio) evaluates to

$$\frac{\partial A}{\partial \iota} = \gamma_\infty \frac{\partial F}{\partial \iota} - \left(\mu - \mu_\iota + \frac{p_\iota - p}{\rho'} \right), \qquad (10)$$

in differential terms, where μ_ι and p_ι are the chemical potential and the pressure inside the droplet, respectively, while μ and p refer to the vapour. In the $\iota \to \infty$ limit, the rightmost term of Eq. (10) yields an effective chemical potential difference

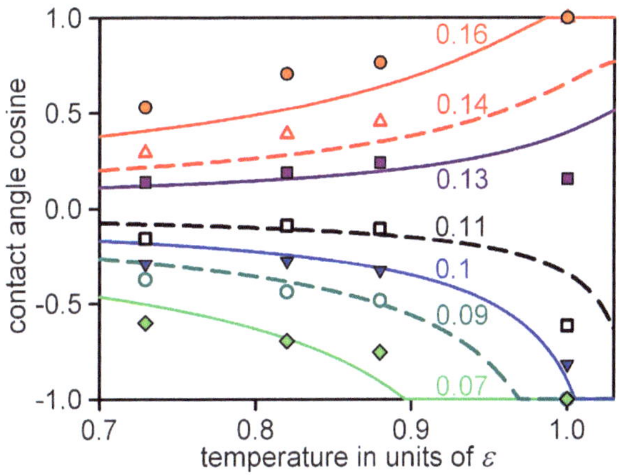

Fig. 5 MD simulation results for ϑ over T with a reduced fluid-wall dispersive energy of $\zeta = 0.07$ (*diamonds*), 0.09 (*circles*), 0.1 (*down triangles*), 0.11 (*open squares*), 0.13 (*black squares*), 0.14 (*triangles*), and 0.16 (*bullets*) as well as following the proportionality law (*lines*), cf. Eq. (7), with $\zeta_0 = 0.118$, $\delta_\infty(T)$ from a correlation based on the data of Vrabec et al. [6], and a reduced fluid-wall surface energy difference of $\mathsf{K}_\gamma = 7\,\sigma\varepsilon$, cf. Eq. (9)

$$\Delta\mu_{\mathrm{e}} = \mu - \mu_{\mathrm{s}}(T) + \frac{p_{\mathrm{s}}(T) - p}{\rho'}, \tag{11}$$

that accounts for the 'pressure effect' [23] of the vapour – which may include the contribution of an inert carrier gas. Note that $\mu_{\mathrm{s}}(T)$ and $p_{\mathrm{s}}(T)$ do not depend on ι since these quantities characterize the saturated bulk fluid. As visualized in Fig. 6, the presence of a carrier gas increases the free energy barrier $\Delta A^\star$ of a nucleation process, i.e. the maximum of ΔA reached for a *critical droplet* in (unstable) equilibrium with the vapour, corresponding to the conditions discussed above.

From the analysis of a random walk over ι, the probability for a droplet containing ℓ molecules to eventually reach macroscopic size can be determined as

$$\mathsf{Q}(\ell) = \frac{\int_1^\ell \exp(2\Delta A/T)\, d\iota}{\int_1^\infty \exp(2\Delta A/T)\, d\iota}, \tag{12}$$

while the nucleation rate, i.e. the number of macroscopic liquid drops formed by homogeneous nucleation per volume and time unit, is

$$J = \mathsf{CNT} \cdot \frac{N' z F^\star}{V} \exp\left(\frac{-\Delta A^\star}{T}\right), \tag{13}$$

according to CNT. In this expression, T refers to the rate at which vapour monomers collide with an interface (per surface area), N' is the number of monomers in the system, $F^\star$ is the surface area of a critical droplet, and V is the system volume. The

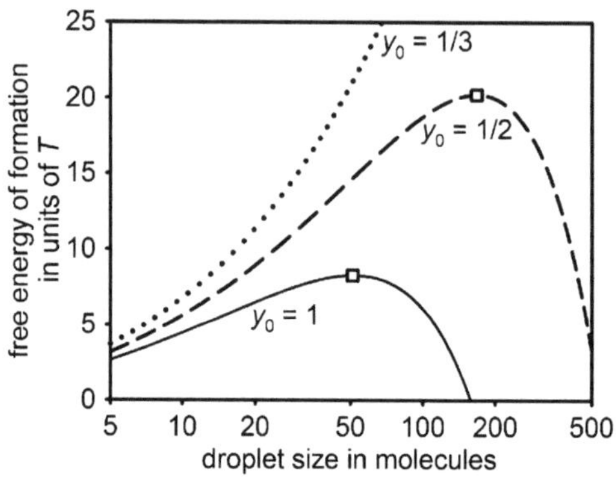

Fig. 6 Free energy of formation according to CNT for CO_2 droplets in supersaturated vapours at $T = 250.2$ K with $S_\rho = 2.72$ and CO_2 mole fractions of $y_0 = 1$ (*continuous lines*), $1/2$ (*dashed lines*), and $1/3$ (*dotted lines*), reaching a maximum for the critical droplet (*squares*). Note that in the thermodynamic limit, i.e. in a macroscopic system, the Gibbs, Helmholtz, and Landau free energies of formation for small droplets converge

correction factors C, N, and z were introduced by Farkas [22], Feder et al. [24], and Zel'dovič [25], respectively. The overall pressure effect on J following Eq. (13), as discussed by Wedekind et al. [23], defines the carrier gas correction factor W.

5 Homogeneous Vapour to Liquid Nucleation: MD Simulation

Nucleation in supersaturated vapours was studied by simulating systems containing between 100,000 and 17,000,000 molecules, exploiting the excellent scalability of the *ls*1 *mardyn* program on massively parallel computing platforms [26].

The method of Yasuoka and Matsumoto [27], where droplet formation rates are evaluated during the stage of a condensation process that corresponds to nucleation (rather than relaxation or droplet growth), was applied to the canonical ensemble. In these simulations, the vapour pressure decreased over time due to the transfer of molecules from the vapour to the dispersed liquid phase. Furthermore, steady state quantities, pertaining to nucleation only, were investigated with a new simulation method. This method combines the grand canonical ensemble with *McDonald's dæmon* [28], an intelligent being whose interventions eliminate the droplets containing more than ℓ molecules; see Fig. 7 for a comparison between these approaches. Results for the LJ·TS fluid (shown in Fig. 8) agree well with CNT, using a temperature independent value for the empirical correction factor C introduced by Farkas [22].

Canonical ensemble MD simulations were also conducted for multi-component systems containing nitrogen, oxygen, and argon – at the ratio prevalent in the earth's atmosphere – as well as carbon dioxide with a greater partial density than

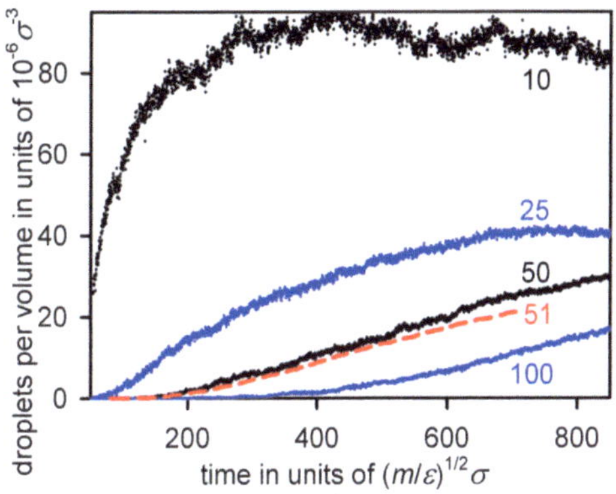

Fig. 7 Number of droplets per volume over simulation time for droplets containing ι > 10, 25, 50, and 100 molecules in a canonical ensemble MD simulation of the LJ·TS fluid at $T = 0.7\ \varepsilon$ and $\rho = 0.03421\ \sigma^{-3}$ in comparison with the aggregated number of dæmon interventions per volume in a grand canonical MD simulation with $T = 0.7\ \varepsilon$, $S_\mu = 2.8658$, and $\ell = 51$

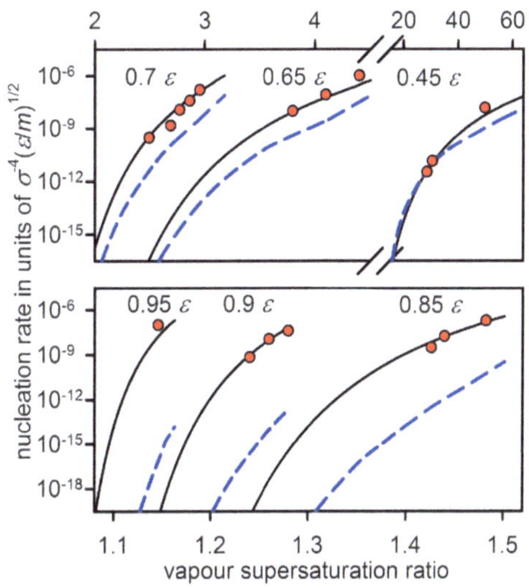

Fig. 8 Nucleation rate of the LJ·TS fluid over supersaturation from the present MD simulations of the grand canonical ensemble with McDonald's dæmon (*black circles*) as well as according to CNT with C = 200 (*dashed lines*) and the Laaksonen et al. [18] model (*dashed dotted lines*) at temperatures of $T = 0.45$, 0.65, 0.7, 0.85, 0.9, and 0.95 ε. The supersaturation ratio is given in terms of the chemical potential of the vapour phase

at saturation. The molecular models employed for this purpose, introduced by Vrabec et al. [29], are well-established with respect to fluid phase equilibria [29–31]. For these systems, the analysis of the carrier gas effect according to Wedekind et al. [23] is confirmed qualitatively by the determined droplet formation rates J_ℓ, given in Table 2, although significant quantitative deviations are present at high temperatures.

6 Conclusion

From the preceding analysis of curved vapour-liquid interfaces and homogeneous nucleation it can be concluded that CNT is able to capture both the nucleation rate and the critical droplet size for the considered systems, i.e. the LJ·TS fluid and a quaternary mixture of quadrupolar and unpolar fluids. The main criticism

Table 2 Droplet formation rate from Yasuoka-Matsumoto (YM) canonical ensemble MD simulation as well as critical droplet size (in molecules), Wedekind factor W, and the prediction $J/\mathsf{Q}_{\mathrm{CNT}}(\ell)$ for the droplet formation rate according to CNT with $\mathsf{C} = 1$, in dependence of temperature (in units of K), supersaturation ratio (with respect to the partial density of carbon dioxide) and YM threshold size ℓ (in molecules) for the quaternary system $\mathsf{CO_2} + \mathsf{N_2} + \mathsf{O_2} + \mathsf{Ar}$. The mole fraction y_0 of carbon dioxide in the supersaturated vapour is indicated in the table, while the composition regarding the other fluids corresponds to the earth's atmosphere. The rates are given in units of $\mathrm{m}^{-3}\mathrm{s}^{-1}$ and where no nucleation was detected, J_{CNT} instead of $J/\mathsf{Q}_{\mathrm{CNT}}$ is shown in the last column

T	ρ/ρ''	y_0	ℓ	J_ℓ	$\iota^\star$	W	$J/\mathsf{Q}_{\mathrm{CNT}}$
238.4	2.80	1/2	50	$1.5 \cdot 10^{33}$	66	0.03	$2.6 \cdot 10^{31}$
			85	$1.6 \cdot 10^{32}$			$3.3 \cdot 10^{30}$
		1	50	$5.6 \cdot 10^{32}$	41	1	$9.9 \cdot 10^{31}$
			85	$2.1 \cdot 10^{32}$			$7.6 \cdot 10^{31}$
	3.08	1/2	50	$5.5 \cdot 10^{33}$	65	0.02	$3.1 \cdot 10^{31}$
			150	$3.1 \cdot 10^{32}$			$3.9 \cdot 10^{30}$
		1	50	$6.3 \cdot 10^{33}$	39	1	$1.6 \cdot 10^{32}$
			150	$2.9 \cdot 10^{32}$			$1.3 \cdot 10^{32}$
	3.36	1/3	—	$\ll 10^{31}$	127	$4.2 \cdot 10^{-6}$	$1.1 \cdot 10^{27}$
		1/2	50	$1.1 \cdot 10^{34}$	65	0.02	$8.7 \cdot 10^{30}$
			300	$3.2 \cdot 10^{32}$			$4.2 \cdot 10^{30}$
		1	50	$6.7 \cdot 10^{33}$	37	1	$2.1 \cdot 10^{32}$
			300	$1.4 \cdot 10^{33}$			$1.8 \cdot 10^{32}$
250.2	2.34	1/2	50	$1.1 \cdot 10^{34}$	140	$1.9 \cdot 10^{-4}$	$1.8 \cdot 10^{33}$
			100	$1.1 \cdot 10^{33}$			$7.8 \cdot 10^{29}$
		1	50	$1.3 \cdot 10^{33}$	54	1	$3.9 \cdot 10^{32}$
			100	$3.4 \cdot 10^{32}$			$1.4 \cdot 10^{32}$
	2.53	1/2	85	$7.4 \cdot 10^{33}$	143	$1.0 \cdot 10^{-4}$	$3.9 \cdot 10^{30}$
			200	$7.4 \cdot 10^{32}$			$3.1 \cdot 10^{28}$
		1	85	$2.2 \cdot 10^{33}$	52	1	$1.9 \cdot 10^{32}$
			200	$7.7 \cdot 10^{32}$			$1.9 \cdot 10^{32}$
	2.72	1/3	—	$\ll 10^{31}$	879	$4.3 \cdot 10^{-25}$	$2.3 \cdot 10^{8}$
		1/2	75	$1.3 \cdot 10^{34}$	150	$4.2 \cdot 10^{-5}$	$1.8 \cdot 10^{31}$
			250	$1.6 \cdot 10^{33}$			$1.7 \cdot 10^{28}$
		1	75	$4.8 \cdot 10^{33}$	50	1	$2.6 \cdot 10^{32}$
			250	$1.4 \cdot 10^{33}$			$2.5 \cdot 10^{32}$

usually made of CNT is that it applies the capillarity approximation to small droplets where significant curvature effects should be expected. However, a deviation from capillarity is implicit in the prefactor C which empirically accounts for its overall influence on the nucleation rate. This corresponds to stating that the capillarity approximation overestimates the free energy barrier by $T \ln \mathsf{C}$.

The physical foundation of this approach is more robust than it might seem at first sight. By combining recent simulation results on the equilibrium vapour pressure of droplets, cf. Fig. 2, it becomes apparent that curvature effects are significant in the immediate vicinity of the spinodal line for the vapour, corresponding to $\iota^\star < 100$, while they are virtually undetectable for droplets containing more than

1,000 molecules. Thus, the deviation from Eq. (10) regarding the magnitude of $\Delta A^\star$ is dominated by an integral over the free energy of formation for *extremely small* droplets. At supersaturation ratios sufficiently distant from spinodal conditions, this contribution does not depend on S and can be represented by $-T \ln \mathsf{C}$ with a constant value of C.

Acknowledgement The authors would like to thank J. Harting, H. Hasse, E.Y. Kenig, and G. Reina for their support and for valuable discussions. The present work, which contributes to the BMBF project IMEMO, was conducted under the auspices of the Boltzmann-Zuse Society of Computational Molecular Engineering (BZS). Major aspects of it were facilitated by the reliable technical assistance of M. Heitzig and Z. Lin. The position of M.T. Horsch at Imperial College London is funded by the DAAD postdoc programme, and computations were performed at the High Performance Computing Center Stuttgart (HLRS) with resources assigned to the grant MMHBF.

References

1. Verschaffelt, J.E.: Bulletin de l'Académie Royale de Belgique: Classe de Sciences **22**(4), 373 (1936)
2. Gibbs, J.W.: American Journal of Science: Series 3 **16**, 441 (1878)
3. Guggenheim, E.A.: Transactions of the Faraday Society **35**, 397 (1940)
4. Tolman, R.C.: Journal of Chemical Physics **17**(3), 333 (1949)
5. Allen, M.P., Tildesley, D.J.: *Computer Simulation of Liquids* (Clarendon, Oxford, 1987)
6. Vrabec, J., Kedia, G.K., Fuchs, G., Hasse, H.: Molecular Physics **104**(9), 1509 (2006)
7. Hołyst, R., Litniewski, M.: Physical Review Letters **100**, 055701 (2008)
8. van Meel, J.A., Page, A.J., Sear, R.P., Frenkel, D.: Journal of Chemical Physics **129**, 204505 (2008)
9. Block, B.J., Das, S.K., Oettel, M., Virnau, P., Binder, K.: Journal of Chemical Physics **133**, 154702 (2010)
10. Horsch, M., Vrabec, J., Hasse, H.: Physical Review E **78**, 011603 (2008)
11. Horsch, M., Vrabec, J.: Journal of Chemical Physics **131**, 184104 (2009)
12. Horsch, M., Heitzig, M., Dan, C., Harting, J., Hasse, H., Vrabec, J.: Langmuir **26**(13), 10913 (2010)
13. Horsch, M., Lin, Z., Windmann, T., Hasse, H., Vrabec, J.: Atmos. Res. **101**(3), 519 (2011)
14. Buff, F.P.: Journal of Chemical Physics **23**(3), 419 (1955)
15. Kondo, S.: Journal of Chemical Physics **25**(4), 662 (1956)
16. Young, T.: Philosophical Transactions of the Royal Society **95**, 65 (1805)
17. Napari, I., Julin, J., Vehkamäki, H.: Journal of Chemical Physics **131**, 244511 (2009)
18. Laaksonen, A., Ford, I.J., Kulmala, M.: Physical Review E **49**(6), 5517 (1994)
19. Bernreuther, M., Vrabec, J.: In: High Performance Computing on Vector Systems, Resch, M. et al. (Eds.) pp. 187–195. Springer, Heidelberg (2006); ISBN 3-540-29124-5
20. ten Wolde, P.R., Frenkel, D.: Journal of Chemical Physics **109**, 9901 (1998)
21. Volmer, M., Weber, A.: Zeitschrift für physikalische Chemie (Leipzig) **119**, 277 (1926)
22. Farkas, L.: Zeitschrift für physikalische Chemie (Leipzig) **125**, 236 (1927)
23. Wedekind, J., Hyvärinen, A.P., Brus, D., Reguera, D.: Physical Review Letters **101**, 125703 (2008)
24. Feder, J., Russell, K.C., Lothe, J., Pound, G.M.: Advances in Physics **15**(1), 111 (1966)
25. Zel'dovič, Â.B.: Žurnal Èksperimental'noj i Teoretičeskoj Fiziki **12**, 525 (1942)
26. Buchholz, M., Bungartz, H.-J., Vrabec, J.: J. Comput. Sci. **2**(2), 124 (2011)
27. Yasuoka, K., Matsumoto, M.: Journal of Chemical Physics **109**(19), 8451 (1998)

28. McDonald, J.E.: American Journal of Physics **31**, 31 (1963)
29. Vrabec, J., Stoll, J., Hasse, H.: Journal of Physical Chemistry B **105**(48), 12126 (2001)
30. Huang, Y.L., Vrabec, J., Hasse, H.: Fluid Phase Equilibria **287**(1), 62 (2009)
31. Vrabec, J., Kedia, G.K., Buchhauser, U., Meyer-Pitroff, R., Hasse, H.: Cryogenics **49**, 72 (2009)

Score-P: A Unified Performance Measurement System for Petascale Applications

Dieter an Mey, Scott Biersdorf, Christian Bischof, Kai Diethelm, Dominic Eschweiler, Michael Gerndt, Andreas Knüpfer, Daniel Lorenz, Allen Malony, Wolfgang E. Nagel, Yury Oleynik, Christian Rössel, Pavel Saviankou, Dirk Schmidl, Sameer Shende, Michael Wagner, Bert Wesarg, and Felix Wolf

D. an Mey · C. Bischof · D. Schmidl
Center for Computing and Communication, RWTH Aachen University, Seffenter Weg 23, 52074 Aachen, Germany
e-mail: anmey@rz.rwth-aachen.de; bischof@rz.rwth-aachen.de; schmidl@rz.rwth-aachen.de

S. Biersdorf · A. Malony · S. Shende
Performance Research Laboratory, University of Oregon, Eugene, OR 97403, USA
e-mail: scottb@cs.uoregon.edu; malony@cs.uoregon.edu; sameer@cs.uoregon.edu

K. Diethelm
GNS Gesellschaft für numerische Simulation mbH, Am Gaußberg 2, 38114 Braunschweig, Germany
e-mail: diethelm@gns-mbh.com

D. Eschweiler · D. Lorenz · C. Rössel · P. Saviankou
Forschungszentrum Jülich GmbH, Jülich Supercomputing Centre, Leo-Brandt-Str., 52428 Jülich, Germany
e-mail: d.eschweiler@fz-juelich.de; d.lorenz@fz-juelich.de; c.roessel@fz-juelich.de; p.saviankou@fz-juelich.de

M. Gerndt · Y. Oleynik
Technische Universität München, Fakultät für Informatik, Boltzmannstraße 3, 85748 Garching, Germany
e-mail: gerndt@in.tum.de; oleynik@in.tum.de

A. Knüpfer (✉) · W.E. Nagel · M. Wagner · B. Wesarg
Technische Universität Dresden, Zentrum für Informationsdienste und Hochleistungsrechnen (ZIH), 01062 Dresden, Germany
e-mail: andreas.knuepfer@tu-dresden.de; wolfgang.nagel@tu-dresden.de; michael.wagner2@tu-dresden.de; bert.wesarg@tu-dresden.de

F. Wolf
Forschungszentrum Jülich GmbH, Jülich Supercomputing Centre, Leo-Brandt-Str., 52428 Jülich, Germany

German Research School for Simulation Sciences, Laboratory for Parallel Programming, Schinkelstr. 2a, 52062 Aachen, Germany

Department of Computer Science, RWTH Aachen University, Ahornstr. 55, 52074 Aachen, Germany
e-mail: f.wolf@grs-sim.de

C. Bischof (eds.), *Competence in High Performance Computing 2010*,
DOI 10.1007/978-3-642-24025-6_8,

Abstract The rapidly growing number of cores on modern supercomputers imposes scalability demands not only on applications but also on the software tools needed for their development. At the same time, increasing application and system complexity makes the optimization of parallel codes more difficult, creating a need for scalable performance-analysis technology with advanced functionality. However, delivering such an expensive technology can hardly be accomplished by single tool developers and requires higher degrees of collaboration within the HPC community. The unified performance-measurement system Score-P is a joint effort of several academic performance-tool builders, funded under the BMBF program *HPC-Software für skalierbare Parallelrechner* in the SILC project (*Skalierbare Infrastruktur zur automatischen Leistungsanalyse paralleler Codes*). It is being developed with the objective of creating a common basis for several complementary optimization tools in the service of enhanced scalability, improved interoperability, and reduced maintenance cost.

1 Introduction

Today, computer simulations play an increasingly critical role in many areas of science and engineering, with applications growing both in number and sophistication. This creates a rising demand for computing capacity, both in terms of the number of systems and in terms of the computational power offered by individual systems. After we can no longer count on the rapid speed improvements of uniprocessors, supercomputer vendors answer this demand today with an increasing number of cores per system, forcing users to employ larger process configurations. Furthermore, modern systems feature hybrid and often also heterogeneous designs with deep memory hierarchies and advanced network architectures, further complicating the programming task. Therefore, performance-analysis tools are essential instruments in the hand of application developers that help them to cope with this complexity and to understand the performance implications of their software design choices. This is in particular true on emerging platforms whose performance characteristics are not yet well understood.

1.1 Scalability Challenge

It is often neglected that many parallel programming tools face scalability challenges, just as the applications they are designed for. In fact, performance tools are most urgently needed when scaling an application to unprecedented levels, for example, in the pursuit of multi-petascale performance. Tools must even have the edge over applications with respect to the number of processes at which they can operate. In an ideal world, tools should always be readily available at the highest available scale.

1.2 Interoperability Challenge

In the past, the authors developed a number of complementary performance tools such as Periscope [6], Scalasca [5], Vampir [8], and TAU [19], each focusing on a different aspects of the performance behavior. Although one would like to use them in combination, this is complicated by the fact that for historic reasons each of them uses a proprietary measurement system with its own set of data formats. Since the data formats are very similar, conversion tools alleviated this in the past. The alternative is re-running the experiment with another tool's measurement system. Both ways are very inconvenient for the users and become more troublesome with increasing scale.

1.3 Redundancy Challenge

Although all four tools follow distinctive approaches and pursue individual strategies on how to address today's demand for performance-analysis solutions, they share certain features and base functionalities. This includes, for example, the instrumentation and measurement modules. Also, the data formats have very similar semantics but slightly different ways of representation. As a consequence, the growing effort required for code maintenance, feature extensions, scalability enhancements, and user support is effectively multiplied.

In this paper, we report on the status and intermediate results of the SILC project, which aims at the design and implementation of a joint measurement infrastructure for supercomputing applications called *Score-P*. The highly scalable and easy-to-use infrastructure will serve as a common basis for the above-mentioned performance tools Periscope, Scalasca, Vampir, and TAU. The project partners look back on a long history of collaboration, in particular through the *Virtual Institute – High Productivity Supercomputing* (VI-HPS) [21], a Helmholtz-funded initiative of academic HPC tool builders from which the idea for this project emerged.

We argue that a joint performance-measurement infrastructure, the part where the overlap between the tools is significant, in combination with common data formats will not only improve interoperability but also notably reduce the overall development cost. Although a joint infrastructure will entail more coordination among developers from previously independent teams and will create more complex dependencies between the common components on the one hand and features of the individual analysis tools on the other hand, we believe that in the end, such a collaboration will save substantial resources that can be better spent on adding new features, further improving the software quality, and providing user support. For example, the savings will open the way for more powerful scalability enhancements of the measurement system alongside more advanced analysis functionality, substantially adding to the overall user value.

The next section introduces the project partners and outlines the project goals. The background and related-work section discusses the tools involved in this project as well as other well-known HPC performance analysis tools. The remainder of the paper, from Sect. 4 to Sect. 9, will discuss key software components, namely the Score-P instrumentation and runtime system, the event trace data format Open Trace Format Version 2 (OTF2), the CUBE4 profile data format, the Online Access (OA) interface, the OpenMP instrumenter Opari2, and the interface to the TAU tools. Finally, there will be an outlook on future work.

2 Project Overview and Goals

The SILC project (*Skalierbare Infrastruktur zur automatischen Leistungsanalyse paralleler Codes*, Engl. scalable infrastructure for automatic performance analysis of parallel codes) is a collaboration between the following partners:

- Center for Computing and Communication, RWTH Aachen,
- Zentrum für Informationsdienste und Hochleistungsrechnen (ZIH), TU Dresden,
- Jülich Supercomputing Centre, Forschungszentrum Jülich GmbH,
- Fakultät für Informatik, Technische Universität München, and
- Gesellschaft für numerische Simulation mbH (GNS), Braunschweig,

coordinated by TU Dresden. Also, the following associated partners are involved:

- Performance Research Laboratory, University of Oregon, Eugene/OR/USA,
- German Research School for Simulation Sciences, Aachen, and
- Gesellschaft für Wissens- und Technologietransfer, TU Dresden.

It was proposed by the consortium in 2008 and is funded under the BMBF call *"HPC-Software für skalierbare Parallelrechner"* from 01/2009 until 12/2011. It is carried out in close cooperation with PRIMA, a joint project between the University of Oregon and Forschungszentrum Jülich funded by the US Department of Energy.

The SILC Project Goals

The SILC project will design and implement the joint measurement infrastructure Score-P as a common basis for the performance tools Periscope, Scalasca, Vampir, and TAU. Score-P will provide the following functional requirements:

- Provide event trace recording and profile collection functionality satisfying the needs of all analysis tools involved.
- Implement direct instrumentation of target applications, as currently provided by the proprietary solutions. We plan to add sampling as an alternative in the future.
- Support postmortem and on-line analysis.
- Focus on target applications using MPI and/or OpenMP. In the future, also heterogeneous platforms with GPGPUs should be supported.
- Integrate all tools with the new infrastructure such that they provide their original functionality on top of it. The individual analysis tools will be continued as separate tools under their own names by their respective groups, however.

In addition, the partners agreed on the following non-functional requirements:

- The software should be portable to all relevant Unix based HPC platforms.
- The new infrastructure has to scale to the peta-scale level, that means to hundreds of thousands of processes or threads.
- The measurement overhead should be minimized to produce as little perturbation as possible in the recorded performance data.
- The Score-P software should reach *production quality*, that means it should be robust and well tested against all expected usage scenarios. Furthermore, we will offer appropriate user documentation, support, and training.
- The release at the end of the funding period will be under a New BSD Open Source License, which allows almost any usage.

All partners are committed to a long-term collaboration to further maintain and enhance the results of the SILC project. After the funding period, the joint effort will be open to other tools groups as new partners. The SILC web page [20] and the Score-P web page [17] provide more information. A pre-release version is already available. Future updates will be announced there.

3 Background and Related Work

Since performance analysis is an important part of today's HPC application development, there are a number of tools with emphasis on different aspects. They use either sophisticated profiling techniques or rely on event trace recording.

Periscope [6] is an online performance analysis tool that characterizes an application's performance properties and quantifies related overheads. *Scalasca* [5] is an automatic performance analysis tool which detects a wide range of performance problems and presents the result in a concise graphical representation. It is especially well-suited for communication and synchronization bottlenecks and is extremely scalable. *Tau* [19] is an open source performance analysis framework which mainly relies on sophisticated profile recording and evaluation methods but also supports event tracing. *Vampir* [8] is an interactive event trace browser which visualizes parallel programs with a number of displays showing different aspects of the performance behavior.

The four above-mentioned tools are the primary beneficiaries of the Score-P measurement system. They will remain separate tools, but closely integrated with Score-P as their common measurement system. This is an unmatched level of integration of tools from different development teams, to the best of our knowledge.

The tools *Paraver* and *Dimemas* [11] developed by the Barcelona Supercomputing Center provide interactive event trace visualization and trace-based replay. They allow performance analysis as well as simulation of parallel run-time behaviour under altered conditions. Also, the *Jumpshot* [1] series of tools by the Argone National Laboratory and the University of Chicago provide event trace visualization in a similar way to Vampir and Paraver. *OpenSpeedShop* [16] is a community project by the Los Alamos, Lawrence Livermore and Sandia National Laboratories, and the Krell Institute. It relies mainly on profiling and sampling but supports also event tracing. The *HPCToolkit* [13] by Rice University has similar goals but uses profile recording combined with binary analysis to obtain insights into parallel performance.

4 The Score-P Measurement System

The Score-P measurement system (see Fig. 1) enables users to instrument C/C++ or Fortran applications with probes that collect performance data when triggered during measurement runs. The data is collected as traces and/or profiles and is passed on to one or more back-ends in order to be analyzed postmortem in OTF2, CUBE4 or TAU snapshot format or by Periscope via the on-line interface.

Score-P supports the programming paradigms *serial*, *OpenMP*, *MPI* and *hybrid* (MPI combined with OpenMP). In order to instrument an application, the user needs to recompile the application using the Score-P instrumentation command, which is added as prefix to the original compile and link lines. It automatically detects the programming paradigm by parsing the original build instructions and utilizes appropriate and configurable methods of instrumentation. These are currently:

- Compiler instrumentation
- MPI library interposition

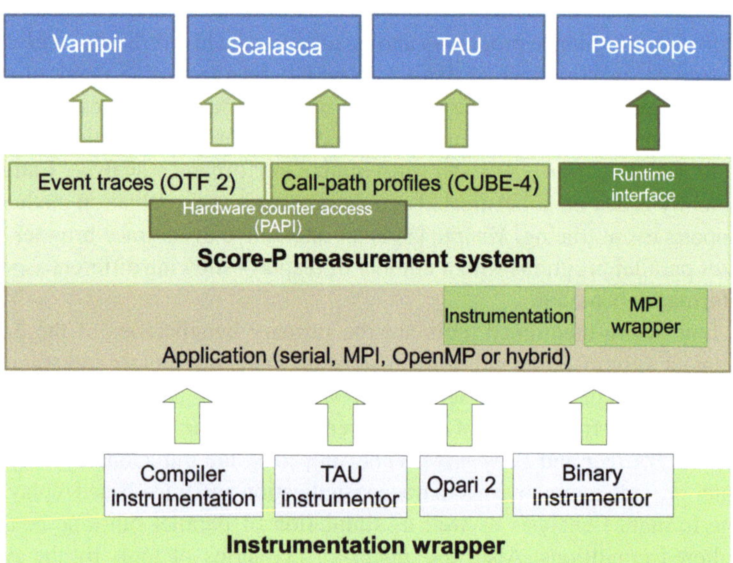

Fig. 1 Overview of the Score-P measurement system architecture and the tools interface

- OpenMP source code instrumentation using Opari2 (see Sect. 8)
- Source code instrumentation via the TAU instrumenter [4]

Additionally, the user may instrument the code manually with convenient macros provided by Score-P. Furthermore, there are ongoing efforts to add instrumentation of executables using binary rewriting. As an alternative to direct instrumentation, we plan to provide sampling functionality in the future.

During measurement, the system records several performance metrics including execution time, communication metrics, and optionally hardware counters. Performance data is stored in appropriately sized chunks of a preallocated memory buffer that are assigned to threads on demand, efficiently utilizing the available memory and avoiding measurement perturbation by flushing the data to disk prematurely.

Without recompilation, measurement runs can switch between *tracing* or *profiling* mode. In tracing mode, the performance events are passed to the OTF2 back-end (see Sect. 5) and are written to files for subsequent postmortem analysis using Scalasca or Vampir. In profiling mode, the performance events are summarized at runtime separately for each call path like in Scalasca. Additionally, we integrated support for phases, dynamic regions and parameter-based profiling known from TAU. The collected data is passed to the CUBE4 back-end (see Sect. 6) for postmortem analysis using Scalasca or TAU or is used directly through the online access interface by Periscope. Also in profiling mode, Score-P supports the automatic detection of MPI wait states. Usually such inefficiencies are important bottlenecks and are thoroughly investigated by means of automatic trace analysis and subsequent visual analysis using a time-line representation. In the case of Score-P wait time profiling, inefficiencies are detected immediately when the respective MPI call is completed and stored as an additional metric in the call-path profile.

5 The Open Trace Format 2

The Open Trace Format 2 (OTF2) is the joint successor of the Open Trace Format (OTF) [7] used by Vampir and the Epilog format [22] used by Scalasca. The new trace format consists of a specification of record types, in conjunction with a new trace writer and reader library. The basic OTF2 record-set is a full merge of the two predecessor formats, retaining their previous features. In the near future, OTF2 will serve as the default data sink for the upcoming Score-P measurement system (see Sect. 4), and as the default data source for the trace-analysis tools Vampir and Scalasca. This enables the user of those tools to analyze the same trace file with multiple tools without the burden of providing the same trace files in different formats. Furthermore, the user is able to combine the advantages of the different analysis tools, e.g., using Vampir to investigate the details of an inefficiency pattern that was previously detected by Scalasca.

The OTF2 library consists of three layers. The first one includes the external API and is responsible for the record representation. The record representation operates directly on the second layer, which is responsible for the memory representation of the trace data. The third layer handles the interaction with the file system and is also responsible for requests from the memory layer. The new trace library includes new features that influence its usage, which are explained below.

The external API layer comes with specifications for MPI 2.0, OpenMP 3.0 and event record types already known from OTF or Epilog. In addition, it is possible to easily add arbitrary data fields to existing records. This will simplify the process of adding new kinds of analysis data, both, for experimental additions or permanent new analysis features in the tools.

In contrast to its predecessors, OTF2 has an internal memory buffer module that is hidden from the programmer and the application using OTF2. This buffer offers several methods of reducing the size of the trace data. Thus, the tools are capable of tracing larger parts of an application without the need to interrupt and perturb the application behavior while flushing the data to disk. The two main techniques to achieve this are runtime run-length compression and support for balancing the available memory between threads of the same process (see Sect. 4).

On the lowest level, the file system interaction layer has a flexible substrate layout to support and easily add different strategies for file writing. OTF2 will support basic compressed (gzip) and uncompressed file writing, as well as more scalable approaches like writing via SIONlib [2]. Because of this flexible substrate layout, new file substrates can be easily plugged in. In addition to writing the data to disk, the complete in-memory data can be directly handed over to another application (e.g. an analysis tool). This will result in much shorter analysis cycles (time for running the measurement and examining the analysis results), because the expensive file-system operations can be skipped.

6 The CUBE Profiling Format Version 4

Designed as a generic file format for representing call-path profiles of parallel programs, CUBE is already supported by a number of HPC programming tools. These include not only Scalasca, for which it has been primarily developed, but also performance profilers like PerfSuite [10], ompP [3], TAU [19] and the MPI error detection tool Marmot [9].

A CUBE file represents summary data from a single program run. Its internal representation follows a data model consisting of three dimensions: metric, call-tree, and system. Motivated by the need to represent performance behavior on different granularity levels and to express natural hierarchical relationships among metrics, call paths, or system resources, each dimension is organized in a hierarchy. CUBE consists of a reader and writer library as well as a free graphical browser (Fig. 2) to interactively explore data files.

While working efficiently for applications with in the order of 10,000 processes, CUBE 3, the current version, which stores all data in a single XML file, starts reaching scalability limits beyond this scale. Major bottlenecks are writing a file to

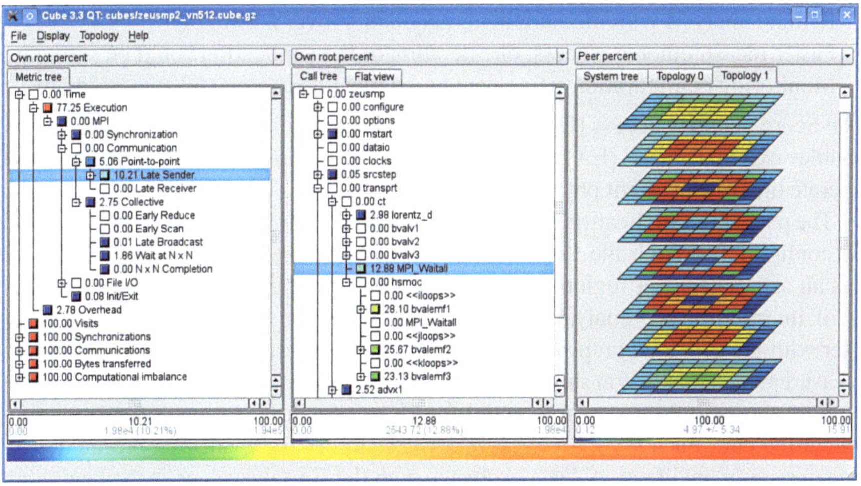

Fig. 2 The CUBE browser displaying the dimensions metric, call path, and system (*left* to *right*)

disk and the footprint of the associated memory representation when loading a file into the browser, seriously impairing interactive user experience. CUBE version 4, which is developed as part of the SILC project, will therefore introduce major changes in the service of enhanced scalability, with the two most important ones listed below:

- To speed up writing data sets, metric values will be stored in a binary format. Human-readable XML will be retained only for the metadata part.
- To reduce the memory footprint of data sets in the browser, the new format will offer random access to individual metrics, which can then be loaded separately. In addition, data for individual call paths will be stored with inclusive semantics, enabling the efficient incremental expansion of the call tree in the browser.

Finally, CUBE 4 will offer a more powerful data model, supporting the representation of time-series and parameter profiles as well as more flexibility in the specification of system-resource hierarchies and display parameters.

7 The Online Access Interface

Another novel feature of Score-P is the possibility to perform measurements in the on-line mode, i.e. control, retrieve and analyze measurements while the application is still running. There are several important benefits which are:

- Reduction of the simultaneously measured/stored performance data
- Possibility for multiple experiments within one run
- Avoiding dumping all measurements to a file at the end

- Remote analysis with measurements acquisition over networks
- Faster measurements process: one iteration of the application could be sufficient
- Monitoring configuration refinement based on already received measurements

The Score-P online access (OA) module, which is part of the measurement system, enables external agents (EA) to connect to the Score-P over TCP/IP sockets and to operate the measurement process remotely.

The part of the application execution for which performance measurements could be configured through the OA interface is called online phase. The online phase has an associated user region containing the part of application source code which is of interest for the analysis and therefore has to be marked manually by the user with the provided preprocessing directives. In order to benefit from multi-step measurements, this region should be an iteratively executed part of the code (e.g. the body of the main loop) with potential for global synchronization at the beginning and at the end. Each phase region will become a root for a call-tree profile during one measurement iteration. Data exchange with the EA takes place at the beginning and at the end of the phase, thus it does not affect the measurements within the phase.

The communication with the EA is done over TCP/IP sockets using a text-based monitoring request interface language which is a simplified subset of the request language used by Periscope. The syntax of the language covers a broad range of online analysis scenarios by means of three kinds of major requests:

- Measurement configuration request
- Execution request
- Measurement retrieval request

The first category of requests allows enabling or disabling of performance metrics available in Score-P. The scope of enabled metrics is global, i.e. they are measured for every region within an online phase. Also some measurement tuning adjustments like depth limits for profile call-trees or filtering of high-overhead regions can be done with these requests. Execution requests are used to control multiple experiments by ordering Score-P to run to the beginning or to the end of the phase or, if the analysis is done, to terminate the application. Measured performance data, stored inside the Score-P call-tree profile, can be accessed by means of measurement retrieval requests. The profile data can be returned to the EA in two ways: as a call-tree profile, where each node represents one call-path of the source code region with associated measurements attached, or as a flat profile, where measurements performed on some source code region are aggregated regardless of the call-path.

8 The Opari2 Instrumenter

Opari [14] is an established source-to-source instrumenter for OpenMP programs which is used in performance tools like Scalasca, VampirTrace and ompP. It automatically wraps OpenMP constructs like parallel regions with calls to the portable OpenMP monitoring interface POMP [14]. In order to support version 3.0 of the OpenMP specification [15], we enhanced Opari to support OpenMP tasking and to provide POMP implementors with information for OpenMP nesting. Furthermore, we improved the usability of the tool itself.

With *tasking*, the OpenMP specification introduced an additional dimension of concurrency. Although this new dimension is convenient, it challenges event-based performance analysis tools because it may disrupt the classic sequence of region entry and exit events. The solution was distinguishing individual task instances and tracking their suspension and resumption points [12].

Traditional performance analysis tools usually pre-allocate memory buffers for a fixed number of threads and store the collected data separately for each thread ID. With OpenMP *nesting* this approach needs to be adjusted because neither is the number of threads known in advance nor is the OpenMP thread ID any longer unique. Therefore, Opari2 provides an upper bound of threads in the next parallel region and an efficient mechanism to access thread-local memory.

During compilation of an application, the previous Opari approach listed all OpenMP constructs in a single file. This was inconvenient for multi-directory project layouts and it prohibited using pre-instrumented libraries or parallel builds. With the new scheme, all relevant OpenMP data stay within the instrumented compilation unit and an enhanced linking procedure offers access to the required data.

All these improvements required interface changes in the POMP specification, which justifies the step from Opari to Opari2 and from POMP to POMP2. With the new version, the established OpenMP instrumentation method is prepared for state-of-the-art parallelization with OpenMP alone or in combination with other methods.

9 Interfacing with TAU

The TAU Performance System® [19] is an open source framework and tools suite for performance instrumentation, measurement, and analysis of scalable parallel applications and systems. TAU provides robust support for observing parallel performance (profiling and tracing) on a broad range of platforms, for managing multi-experiment performance data, and for characterizing performance properties and mining performance features.

TAU provides comprehensive performance instrumentation capabilities that support pre-processor based instrumentation implemented with the *tau_instrumentor* [4], compiler based instrumentation, MPI, POSIX I/O, CUDA, and OpenCL wrapper interposition library based on linking and pre-loading, a binary rewriter (*tau_run* based on DyninstAPI [18]), as well as Python and Java based interpreter level instrumentation implemented with JVMTI.

TAU's instrumentation interfaces with the Score-P measurement library via a special TAU adapter. TAU instrumentation can thus directly layer upon Score-P efficiently by creating a one-to-one mapping between the TAU and Scorc-P profiling constructs. When TAU is configured to use Score-P, it uses Score-P's MPI wrapper interposition library too. TAU's internal data structures are based on tables while Score-P has a tree based storage. This is more natural and efficient for implementing callpath profiling and further reduces TAU's measurement overhead. Using Score-P TAU can generate OTF2 traces that are unified and may be loaded in analysis tools (e.g., Vampir) without having to merge or convert trace files. The online unification of local to global event identifier also removes the need to rewrite the binary traces and the analysis stage can begin immediately after the program completes.

10 Future Work

In the remainder of the funding period, the first official release of Score-P will be prepared. This includes the completion of all scheduled features as well as quality improvements to provide a fully-functional production-quality software package under an Open Source license towards the end of 2011.

After expiration of the funding period, all partners are committed to continuing the joint development and maintenance of Score-P. This will also include user training as part of the dissemination plan. Furthermore, we plan to add new features in the mid-term future. This includes, e.g., a light-weight version of the measurement system that is suitable for permanent performance monitoring or support for accelerator architectures like GPUs with CUDA and OpenCL. Also, new analysis functionality is planned on top of the Score-P system. This may require extensions, additional data items to be collected, or online pre-processing or pre-analysis within the measurement system. Finally, at the end of the funding period, our consortium will be open to new partners who want to attach their tools to Score-P.

References

1. Chan, A., Ashton, D., Lusk, R., Gropp, W.: Jumpshot-4 Users Guide. Mathematics and Computer Science Division, Argonne National Laboratory (2007). ftp://ftp.mcs.anl.gov/pub/mpi/slog2/js4-usersguide.pdf
2. Frings, W., Wolf, F., Petkov, V.: Scalable Massively Parallel I/O to Task-Local Files. In: Proc. of the ACM/IEEE Conf. on Supercomputing, pp. 1–11 (2009)
3. Fürlinger, K., Moore, S.: OpenMP-centric Performance Analysis of Hybrid Applications. In: Proc. of the 2008 IEEE Int. Conf. on Cluster Computing, pp. 160–166. Tsukuba (2008)
4. Geimer, M., Shende, S.S., Malony, A.D., Wolf, F.: A Generic and Configurable Source-Code Instrumentation Component. In: ICCS 2009: Proc. of the 9th Int. Conf. on Computational Science, pp. 696–705. Springer, Berlin (2009)
5. Geimer, M., Wolf, F., Wylie, B.J., Ábrahám, E., Becker, D., Mohr, B.: The Scalasca Performance Toolset Architecture. Concurrency and Computation: Practice and Experience **22**(6), 702–719 (2010)
6. Gerndt, M., Fürlinger, K., Kereku, E.: Periscope: Advanced Techniques for Performance Analysis. In: Parallel Computing: Current & Future Issues of High-End Computing, Proc. of the Int. Conf. ParCo 2005, *NIC Series*, vol. 33, pp. 15–26. Forschungszentrum Jülich (2006)

7. Knüpfer, A., Brendel, R., Brunst, H., Mix, H., Nagel, W.E.: Introducing the Open Trace Format (OTF). In: Computational Science – ICCS 2006, *LNCS*, vol. 3992, pp. 526–533. Springer, Berlin (2006)
8. Knüpfer, A., Brunst, H., Doleschal, J., Jurenz, M., Lieber, M., Mickler, H., Müller, M.S., Nagel, W.E.: The Vampir Performance Analysis Tool Set. In: Tools for High Performance Computing, pp. 139–155. Springer, Berlin (2008)
9. Krammer, B., Müller, M.S., Resch, M.M.: Runtime Checking of MPI Applications with MARMOT. In: Proc. of Parallel Computing (ParCo), pp. 893–900. Málaga (2005)
10. Kufrin, R.: PerfSuite: An Accessible, Open Source Performance Analysis Environment for Linux Development and Performance. In: 6th Int. Conf. on Linux Clusters: The HPC Revolution. Chapel Hill, NC (2005)
11. Labarta, J., Girona, S., Pillet, V., Cortes, T., Gregoris, L.: DiP: A Parallel Program Development Environment. In: Proc. of 2nd Int. EuroPar Conf. (EuroPar 96). Lyon (1996)
12. Lorenz, D., Mohr, B., Rössel, C., Schmidl, D., Wolf, F.: How to Reconcile Event-Based Performance Analysis with Tasking in OpenMP. In: Proc. of 6th Int. Workshop of OpenMP (IWOMP), *LNCS*, vol. 6132, pp. 109–121. Springer, Berlin (2010)
13. Mellor-Crummey, J., Fowler, R., Marin, G., Tallent, N.: HPCView: A tool for top-down analysis of node performance. J. Supercomput. **23**(1), 81–104 (2002)
14. Mohr, B., Malony, A.D., Shende, S., Wolf, F.: Design and Prototype of a Performance Tool Interface for OpenMP. J. Supercomput. **23**(1), 105–128 (2002)
15. OpenMP Architecture Review Board: OpenMP Application Program Interface, Version 3.0. http://www.openmp.org/mp-documents/spec30.pdf, May 2008
16. Schulz, M., Galarowicz, J., Maghrak, D., Hachfeld, W., Montoya, D., Cranford, S.: Open|SpeedShop: An Open Source Infrastructure for Parallel Performance Analysis. Scientific Programming **16**(2-3), 105–121 (2008)
17. Score-P project page. http://www.score-p.org (2010)
18. Shende, S., Malony, A., Morris, A.: Improving the Scalability of Performance Evaluation Tools. In: Proc. of the PARA 2010 Conf. (2010)
19. Shende, S.S., Malony, A.D.: The TAU Parallel Performance System. International Journal of High Performance Computing Applications **20**(2), 287–311 (2006)
20. SILC project page. http://www.vi-hps.org/projects/silc (2009)
21. VI-HPS project page. http://www.vi-hps.org (2010)
22. Wolf, F., Mohr, B.: EPILOG Binary Trace-Data Format. Tech. rep., Forschungzentrum Jülich (2005)

Parallelization of an Edge- and Coherence-Enhancing Anisotropic Diffusion Filter with a Distributed Memory Approach Based on GPI

Martin Kühn

Abstract Numerical algorithms in the seismic industry are among the most challenging areas of High Performance Computing and require an ever growing number of computing power and main memory. The Global Address Space Programming Interface (GPI) provides a model to program distributed memory clusters based on RDMA transfers in a Partitioned Global Address Space (PGAS). Based on GPI a generic straight forward parallelization of an Anisotropic Diffusion Filter (ADF) is implemented as an example of an Explicit Finite Difference scheme. Key features of the implementation are a complete overlay of the computation with network data transfers, a dynamic load distribution scheme and the usage of one-sided communication patterns throughout the algorithm to orchestrate read and write access to the image data. Synchronization points between the compute nodes or barriers are completely avoided. Benchmarks on a cluster with 260 nodes and 1040 cores reveal a constant communication overhead of less than 6% of the total computation time. This figure is still met if the compute nodes in the cluster differ significantly in performance capacity.

1 Introduction

In the past five years the ever growing demand for computing power in the seismic field as in many other fields of scientific computing has been met mainly by increasing the number of CPUs in large computing clusters. Today even small clusters reach handily a triple-digit number of computing cores by employing modern multi-core CPUs. The predominant programming model on these clusters is the Message Passing Interface (MPI) standard. However, with growing numbers of

M. Kühn (✉)
Fraunhofer ITWM, Fraunhofer-Platz 1, 67663 Kaiserslautern, Germany

C. Bischof (eds.), *Competence in High Performance Computing 2010*,
DOI 10.1007/978-3-642-24025-6_9,

nodes and cores the MPI standard doesn't always seem to be the best programming model for scientific computing problems.

An alternative model is offered by the Global Address Space Programming Interface (GPI) [2]. GPI provides a communication library and a runtime system for high performance parallel applications running on cluster systems. It provides a Partitioned Global Address Space (PGAS) to the application which in turn has direct access to a remote data location via Remote Direct Memory Access (RDMA). This allows an easy and efficient use of one sided communication patterns. Additional features are synchronization primitives such as fast barriers or global atomic counters, which allow the development of parallel programs for large scale applications.

Numerical algorithms in the seismic industry are among the most challenging areas of High Performance Computing. A recurrent task in seismic data processing is smoothing and noise reduction. The Anisotropic Diffusion Filtering (ADF) is a proven concept to build edge-enhancing filters to preserve the typical layer structure of seismic 3D image data. The seismic visualization, processing and analysis tool Pre-Stack Pro (PSPro) includes an Edge and Coherence-Enhancing Anisotropic Diffusion filter (ECED). It is based on an explicit finite difference stencil discretization scheme on a regular grid.

The processing of large images in the triple-digit gigabyte range require a large number of cores and a considerable amount of RAM. However, large numbers of cores and nodes quite naturally increase performance inhomogeneities of any kind inside the computing machine. To cope with these inhomogeneities becomes a more and more important task to yield a well performing and especially a well scaling code. A dynamic load balance scheme is a cornerstone strategy to reach this goal.

The one-sided communication paradigm provided by the GPI programming model fits very well with the efficient and straight forward implementation of such dynamic load balance schemes. As shown in this work it allows to implement communication and dynamic load balancing patterns without the need for synchronization points that are usually introduced by the popular send/receive patterns based on the MPI standard.

2 Algorithmic Details

2.1 *Anisotropic Diffusion Filter*

To understand the structure of the underlying algorithm a few details are sketched here. For more detailed descriptions refer to [3–5]. The seismic 3D images $f : \Omega \rightarrow \mathbb{R}$ are usually scalar-valued represented as single precision floating point data on a cuboid sub domain $\Omega \subset \mathbb{R}^3$. The filtered version $u : \Omega \times [0, \inf[\rightarrow \mathbb{R}$ is computed by regarding f as the initial state of the diffusion equation

$$\partial_t u = \nabla^T (D(J_\rho(\nabla u_\sigma))\nabla u) \tag{1}$$

with homogeneous Neumann boundary conditions. u_σ is defined as

$$u_\sigma(x,t) := (K_\sigma * \tilde{u}(.,t))(x) \tag{2}$$

where $\tilde{u}$ denotes an extension of u from Ω to $\mathbb{R}^3$ e.g. obtained by mirroring and $K_\sigma * \tilde{u}$ denotes the convolution of $\tilde{u}$ with a Gaussian kernel with standard deviation σ. The diffusion Tensor D is a positive semidefinite 3×3 matrix that is defined as a function of the so-called structure tensor

$$J_\rho(\nabla u_\sigma) = K_\rho * \left(\nabla u_\sigma \nabla u_\sigma^T\right) \tag{3}$$

Here also $\nabla u_\sigma \nabla u_\sigma^T$ is convoluted with a Gaussian kernel with standard deviation ρ. To derive $D(J)$ the matrix J is diagonalized. $D(J)$ is derived by adjusting the eigenvalues μ_1, μ_2, μ_3 of J to the eigenvalues $\lambda_1, \lambda_2, \lambda_3$ being

$$\lambda_1 := \begin{cases} 1 & : \quad \mu_1 \le 0 \\ 1 - \exp\left(\dfrac{-C_m}{(\mu_1/\lambda)^m}\right) & : \quad \mu_1 > 0 \end{cases} \tag{4}$$

and $\lambda_2 := \lambda_3 := 1$ with suitable parameters m, C_m and λ.

The calculation of u is performed via an explicit finite differences stencil scheme on a regular grid. As the diffusion tensor D depends on u it is recalculated in each time step. This includes mainly the convolution of u with a Gaussian kernel, calculation of the structure tensor J, convolution of J with a Gaussian kernel and diagonalization of the structure tensor.

2.2 *Parallelization Strategy*

Starting from a single threaded reference implementation of the PSPro ECED filter provided by Joachim Weickert, University of Saarbrücken, a performance optimized implementation is prepared. In a first step the single thread performance is increased by using SSE operations to vectorize the matrix diagonalization part and the Gaussian convolution part. In a second step the code is extended to a multi threaded SMP version. The parallelization scheme applied here constitutes a straight forward static load distribution. It is based on a domain decomposition by cutting the 3D volume in p equally sized slices with p being the number of available cores. The result is a well scaling, well performing SMP code as is shown in Fig. 1.

The main memory demand of this algorithm is about 8 times the image size itself. The main part of the memory consumption is caused by the storage of the six components of the diffusion tensor per voxel.

However, although the SMP implementation is well scaling it is not suitable for many data sets that have to be filtered in the seismic area. Suppose a data set of

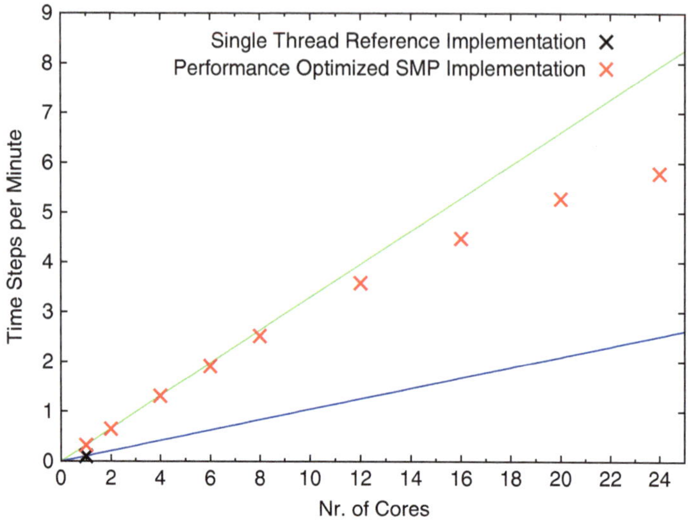

Fig. 1 Benchmark ECED filter, data set $761 \times 512 \times 512$ voxels (4 byte per voxel), green and blue line depict performance under ideal scalability, machine: SMP 4× Intel Xeon E7450 @ 2.4GHz (24 cores), 64 GB RAM

$3{,}500 \times 6{,}000 \times 1{,}251$ voxels has to be filtered in a reasonable time, a SMP machine with around 800 GB RAM and a triple-digit number of cores would be needed which is beyond the equipment of most computing clusters available these days. So a memory distributed approach is necessary to filter these kinds of data sets. The main design objectives of the algorithm are to keep the number of synchronization points as low as possible on one hand and the construction of the algorithm as simple as possible on the other hand.

In the past nonlinear diffusion filters have been implemented and parallelized for distributed memory machines, e.g. in [1]. In that work an isotropic nonlinear diffusion filter is parallelized by distributing the image data among the cluster nodes using two alternating slice partitions. The data set is converted between these two partitions at least once per time step back and forth in an all to all communication pattern. However, the numerical algorithm differs in a few details from the one employed here, e.g. the isotropic diffusion tensor consists of one component only which is not convoluted with a Gaussian kernel.

The advantages of this approach are an easy implementation and that there is no need to make any assumptions on the filter parameters, especially the filter lengths of the Gaussian convolutions.

However, for our application this approach comprises a few important disadvantages. First, it implies a very close synchronization of all the processors in the machine at several points within each time step. As we are aiming for a three to four digits number of processors this is a drawback, as the slowest processor always enforces the total performance between each and every pair of synchronization points. Unfortunately, the performance of all the processors in a compute cluster is not that homogeneous. One reason could be inhomogeneous hardware equipment either because of small deviations in series production or higher deviations because of inhomogeneous CPU or memory equipment of the nodes. Another reason would be on the software side, e.g. the operating system consuming CPU power for an instant on one node but on none of the others etc. Another disadvantage would be that the extensive structure tensor had to be transformed between the two alternating slice partitions each time step as it is too convoluted with a Gaussian kernel. This would generate a demand for high network throughput. A third disadvantage would be that the number of available slices can easily get in the order of magnitude of the number of processors which leads to an inhomogeneous load balance. Consider e.g. the data set example mentioned above and a cluster with 1,000 cores, so we would end up with two partitions where the slices are 3.5 and 6 voxels thick respectively. As 3.5 means that some processors calculate a slice of $3 \times 6{,}000 \times 1{,}251$ voxels, while others calculate slices of $4 \times 6{,}000 \times 1{,}251$ voxels, the load imbalance is evident. The low count of voxels in one single direction also practically prevents the implementation of a dynamic load balance scheme with all the consequences already derived earlier from inhomogeneous performance of the processors.

For these reasons another approach is chosen. Because of the stencil scheme each voxel of the current time step depends only on the same pixel and its neighbors of the previous time step. The same holds for the structure tensor. However, the convolution with the Gaussian kernels increases their dependency range. As the convolution is done with a cutoff, the dependency range in grid points is about $n_d = 1 + l_\sigma + l_\rho$, with l_σ and l_ρ being the cutoff length of the filters. To filter seismic images, a n_d of 10 to 20 grid points is commonly used. Having these additional assumptions in mind the x/y surface of the image data set consisting of $n_x \times n_y \times n_z$ voxels is decomposed into tiles which further on decompose the whole data set into rods of depth n_z. As n_z is very often in the order of $n_z \approx 1{,}000$ grid points there is no need for a subdivision of the image data in z direction.

Contrary to frequently applied strategies, we are not aiming at having compact subset of tiles per node with the objective of minimizing contact surfaces and network transfers. Instead all the tiles are transferred to their processing node via network per default. If the network transfers overlap completely with computation, the increased data traffic is not affecting the latency at all, provided that sufficient network throughput is available. To encourage high network throughput, the transfers are balanced by fine grained tile distributions among the nodes. The reward of this approach is simplicity paired with a very high flexibility in distributing the computational load.

To calculate the current time step of one tile it is sufficient to have the result of the previous time step of that tile itself and a boundary of size n_d around the tile. So a natural working unit of the algorithm is to load a tile including boundary of the previous time step, perform a single time step on this tile including boundary and write back the result without boundary at the same place. However, as the boundary of one tile overlaps with other tiles, it is necessary to load all the tiles around a certain tile before it is written back (compare Fig. 2). To accomplish this, a local

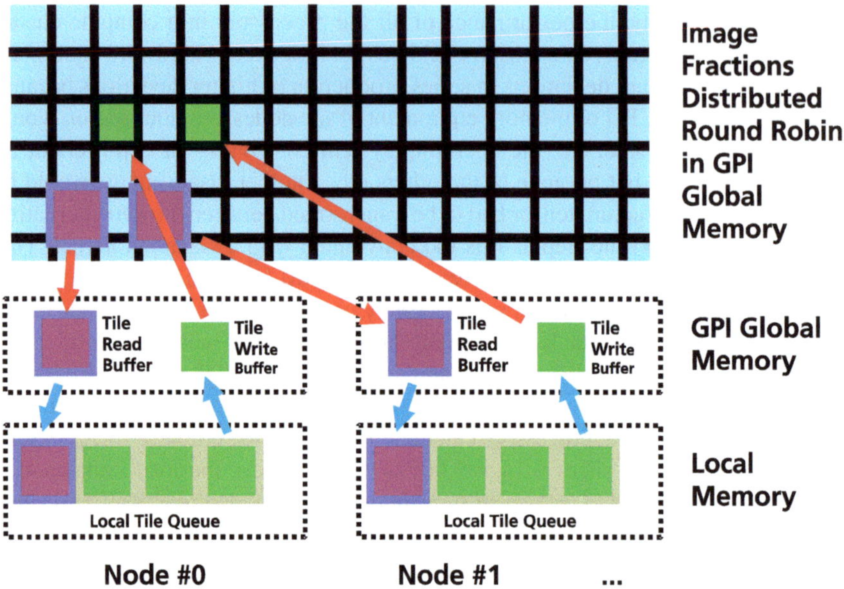

Fig. 2 Snapshot of the data flow during one turn of the algorithm, purple/green tiles belong to the current/next time step respectively, boundaries around the tiles of current time step are marked in dark purple color, red arrows depict RDMA transfers, blue arrows depict memory copies

tile queue is installed for each worker where it can store processed tiles temporarily. Further it is necessary to communicate among all the nodes, which tiles are ready for reading and which tiles are ready for writing.

Obviously this algorithm has an unavoidable need for synchronization because a voxel of the current time step has to be calculated before the same can be calculated again in the next time step. This is inherent for all iterative algorithms. However, the idea is to perform all the necessary communication in a one sided way in combination with a dynamic load balance scheme. In this manner an algorithm is built where in practice no worker has to wait for any synchronization except at the very beginning and the very end of the whole algorithm. However we'll have to make the assumption that there are enough tiles per time step that are calculated on average by each working unit on one hand and that the performance variations between different workers are limited on the other hand.

To choose the size of the tiles one has to consider two criteria. On one hand a high ratio of tiles to workers is favorable, on the other hand a high ratio of tile edge lengths to the dependency range n_d is sought. As n_d is a constant integer depending on user parameters only, there is obviously a tradeoff between these two criteria. To ensure the best possible compromise the workers are not chosen to be a single core, like it is often done in MPI parallel programs, but a single SMP node of the cluster. Each tile is processed on the nodes using the SMP parallel implementation of the ECED filter enhancing the modularity of the final code.

2.3 Communication Patterns

The GPI interface provides functions for one sided communication on a Partitioned Global Address Space (PGAS) which are effectively implemented and straight forward to use. Although one sided communication is also available with the popular MPI-2 standard, it does not seem to be widely applied by many users of the MPI standard. The GPI interface however strongly encourages the use of one sided communication and the overlay of data transfers with computing parts of the algorithm.

Figure 2 depicts a snapshot of the data flow of the algorithm. The image data is stored in the global address space distributed in y/z planes of thickness one grid point in a round robin scheme among the nodes. Further each node reserves two buffers in the global address space located in its own part of the PGAS to collect and store a single tile via RDMA. Finally each node constructs a queue outside the PGAS to store a few tiles temporarily. Although it is not mandatory to place the local tile queues outside the PGAS it helps to save PGAS space at the price of memory copies inside the nodes.

Data transfer and computation proceeds in a turn-based pattern as illustrated in Fig. 2. First, a tile from the end of the local tile queue is stored in the local tile write buffer. A RDMA write is initiated on this tile to store it back in the image data if it is writable. Second, the current tile is imported from the tile read buffer to the local tile queue. Then a RDMA read is initiated on the empty buffer to fetch the next tile from the image data. Third, a time step is performed on the current tile. During this step the data transfers take place without involvement of the CPUs overlapping completely with the computation. As all the RDMA transfers are handled by the network card, they don't cause any performance imbalances among the cores. After the calculation is finished all the open RDMA transfers are checked for completeness. After this the scheme starts over with step one.

Vital to this scheme are the communication patterns that distribute the information about the status of each tile and assign the available work units. The dynamic distribution of the tiles among the nodes is straight forward, as the GPI interface provides several global counters. One of these is reset in the very beginning and incremented by each node asking for assignment of a new tile. The tiles are numbered line by line.

The status communication of each tile is a little more effort. Available for reading in the current time step is any tile that has been written together with all its neighbors in the last time step. A tile available for writing in the current time step is a tile which neighbors have been read completely in the current time step. Now we introduce the following auxiliary variables M_r and M_w

$$M_w = \min\{n_i \in \mathbb{N} | \text{tile } n_i \text{ will be read next by node i}\} \tag{5}$$

$$M_r = \min\{n_i \in \mathbb{N} | \text{tile } n_i \text{ will be written next by node i}\}. \tag{6}$$

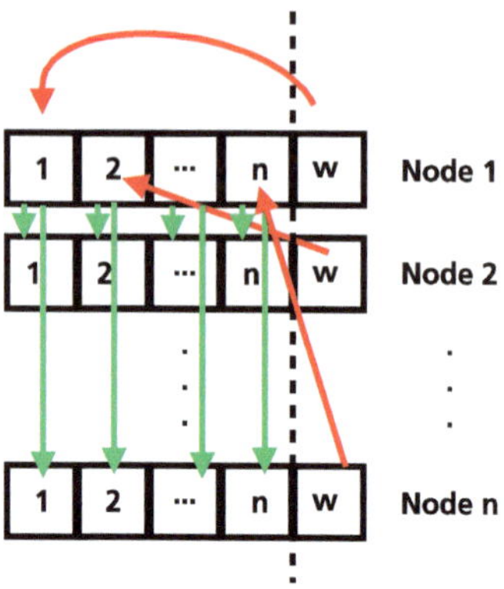

Fig. 3 Collective operation to generate $\tilde{M}_w^i$ and $\tilde{M}_r^i$ on each node $i \in \{1, \ldots, n\}$, red arrows depict write RDMA transfers triggered by the outgoing node, green arrows depict read RDMA operations triggered by the incoming node (polling), all RDMA transfers are triggered once per turn, polling is done by a single RDMA instruction per node and turn

Without loss of generality we assume that n_d is less or equal than the length and width of the tiles, i. e. the boundary is covered by a single neighboring tile in each direction. If we define T_x as the number of tiles per row, it is easy to show that tiles of index $i_T < (M_w - T_x - 1)$ are writable and tiles of index $i_T < (M_r - T_x - 1)$ are readable for next time step. It is important to note that it is not necessary to have the exact value of M_r and M_w at once on all nodes. It is rather sufficient to have estimates $\tilde{M}_r^i \leq M_r$ and $\tilde{M}_w^i \leq M_w$ on each node $i \in \{1, ..., n\}$ which approach at M_r and M_w at some time. It's not even necessary to demand $\tilde{M}_r^i = \tilde{M}_r^j$ for any pair of nodes (i, j).

To generate and distribute these estimates to the nodes a structure is used as depicted in Fig. 3. Each node defines an array of $n + 1$ integer values in its own part of the PGAS. On the index $n + 1$ of that array each node writes e.g. the tile index it will write next. Once per turn in the main algorithm each node i will start a RDMA write from index n+1 of it's own array to index i in the array of the master node 1. On the other hand each node i will poll on the indices $1, \ldots, n$ of the master node by starting a RDMA read on the block ranging from index $1, \ldots, n$ on the master to the same indices of its own array. After the RDMA read is finished, each node i yields an estimate $\tilde{M}_r^i$ by generating a minimum on the indices $1, \ldots, n$ of its own array. As the tile indices that are processed on each node increase monotonously during the algorithm, the condition $\tilde{M}_r^i \leq M_r$ is always met on all nodes i and the estimate will approach at M_r unavoidably.

3 Results

The following benchmarks are performed on the compute cluster "hercules" of the Fraunhofer ITWM. The cluster has 260 nodes equipped with 2 × Intel Xeon 5148 (4 cores), 8 GB RAM, and an Infiniband network on each node.

Figure 4 shows the efficiency of the algorithm which is executed on different number of nodes and different data sets. The plots show the ratio r of the time the

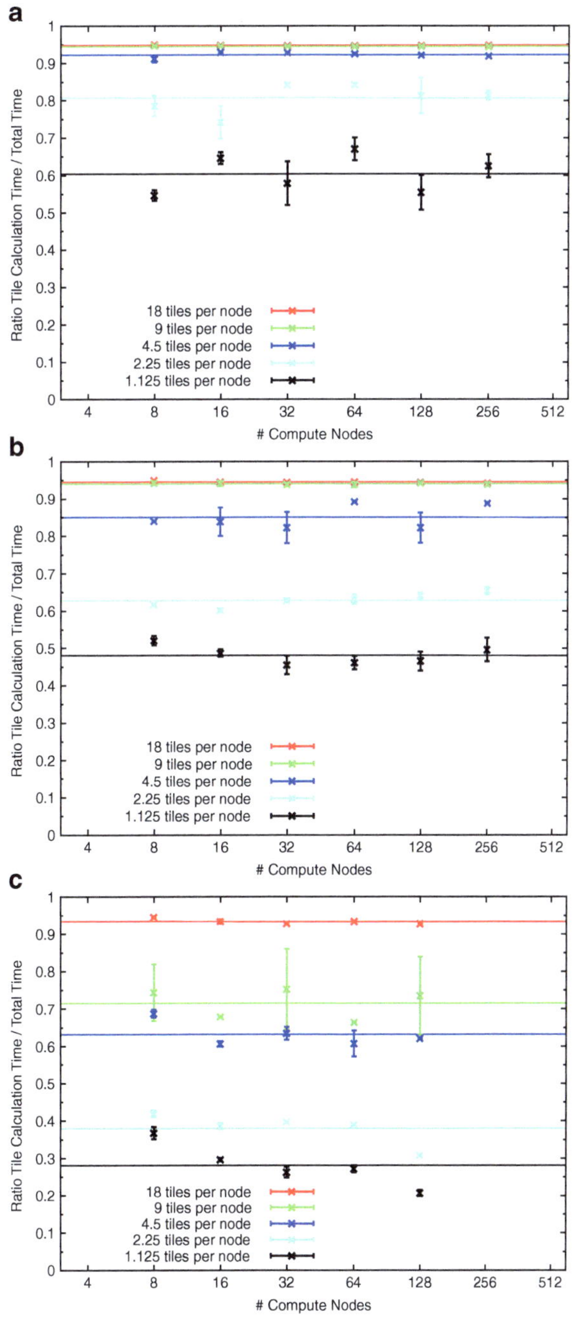

Fig. 4 ECED filter, 10 time steps, tile size $125 \times 125 \times 2{,}400$, dependency range n_d 10 grid points, the horizontal lines are the result of a linear regression with slope zero, data sets: $375 \times 375 \times 2{,}400$, $375 \times 750 \times 2{,}400$, $750 \times 750 \times 2{,}400$, $750 \times 1{,}500 \times 2{,}400$, $1{,}500 \times 1{,}500 \times 2{,}400$, $1{,}500 \times 3{,}000 \times 2{,}400$, $3{,}000 \times 3{,}000 \times 2{,}400$, $3{,}000 \times 6{,}000 \times 2{,}400$, $6{,}000 \times 6{,}000 \times 2{,}400$, $6{,}000 \times 12{,}000 \times 2{,}400$ voxels (4 byte per voxel), averages and standard deviations over five measurements each are presented, machine: cluster hercules (ITWM), (**a**) all nodes use all their cores (**b**) all nodes use all their cores except one node using only two of its cores (**c**) all nodes use all their cores except one node using only on of its cores

nodes spend on calculating the tiles relative to the total time the algorithm takes to finish 10 time steps. The time fraction the algorithm spends on communicating, waiting and copying tiles inside the nodes is given by $1 - r$.

The depth of all the data sets in z direction is 2,400 grid points. The tile size used for all benchmarks is $125 \times 125 \times 2{,}400$ voxel. The data sets are constructed in a way, that the number of grid points in x and y direction is a multiple of the tile size. The dependency range n_d is 10 grid points.

Panel (a) of Fig. 4 demonstrates the result obtained with a homogeneous computing machine. In cases where the number of tiles is 18 or 9 times the number of nodes, a maximum efficiency of 94% is reached. When the ratio of tiles per node drops to 4.5, 2.25 and 1.125, the efficiency drops to 92%, 81% and 60% respectively.

Panel (b) of Fig. 4 demonstrates the result obtained with a inhomogeneous machine. This is simulated by reducing the number of used cores to 2 cores for a single node while all the other nodes continue to use all their 4 cores. This means, that one node has only about 54% the performance of all the other nodes. However, the total efficiency stays at 94% in the cases of 18 and 9 tiles per node, in all the other cases the efficiency drops compared to panel (a).

Panel (c) of Fig. 4 demonstrates the result obtained with an even more inhomogeneous machine. This time the last node uses only 1 core while all the other nodes use all their cores. This means that one nodes has only 27% the performance of all the other nodes. In case of 18 tiles per node the efficiency again stays at 94%. In cases of 9, 4.5, 2.5 and 1.25 tiles per node the efficiency drops compared with panel (b).

Figure 5 demonstrates the yielded speedup on the same cluster for different numbers of nodes. This time an absolute performance number is given in filtered voxels per second. The grid dimensions of the filtered images are chosen to be no multiples of the tile length or width to emphasize the relevance for real live applications. The total computation time of the 256 GB image on 256 nodes (1,024 cores) is approximately 12 minutes for 10 time steps.

The performance value for one compute node in Fig. 5 is calculated with the SMP parallel implementation. As no tile distribution is necessary on a single node, there is no overhead originated in the double computation of the overlapping parts of the tiles. The performance loss because of this overhead p_d for all the other cases is calculated approximately as

$$p_d \approx \frac{n_x^t n_y^t}{(n_x^t + 2n_d)(n_y^t + 2n_d)} \tag{7}$$

with $n_x^t \times n_y^t$ being the size of the tile. In our benchmarks p_d would be $p_d \approx 0.75$. However, the overhead varies from case to case with the filter lengths, size of the image data etc.

Apart from the performance loss p_d the distributed memory implementation proves perfect weak scaling from 2 to 256 nodes which corresponds to 8 to 1,024 compute cores. Further good strong scaling is achieved over one order of magnitude of nodes.

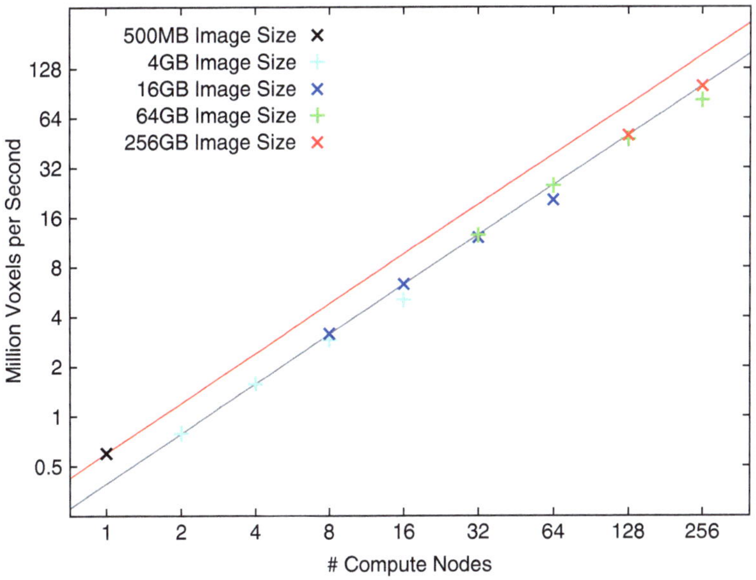

Fig. 5 ECED filter, SMP implementation for 1 node, distributed memory implementation for 2-256 nodes, 10 time steps, tile size 125 × 125 × 2,400, dependency range n_d 10 grid points in x/y direction, data sets: 236 × 236 × 2,400, 668 × 668 × 2,400, 1,337 × 1,337 × 2,400, 2,675 × 2,675 × 2,400, 5,350 × 5,350 × 2,400 voxels (4 byte per voxel), averages over five measurements per point are presented, standard deviations are between 0.1% and 2%, machine: cluster hercules (ITWM)

4 Summary and Conclusion

The goal of this paper is to demonstrate the possibilities offered by the GPI programming model to generate very well scaling codes on hundreds and thousands of CPU's. The one-sided communication patterns realized by RDMA transfers on a Partitioned Global Address Space (PGAS) offer straight-forward implementations of dynamic load balance schemes and the complete overlay of communication with computation.

An Anisotropic Diffusion Filter on 3D seismic image data is chosen as an example of a Finite Difference Stencil code. Image data sets in the seismic area can reach easily hundreds of gigabytes in size and thus require a high number of compute cores and a considerable amount of RAM to be processed in a reasonable time.

Our parallelization approach on a distributed memory machine is based on a domain decomposition of the data set in 2D tiles which are processed independently during each time step. The tile assignment to the compute nodes is done dynamically based on a global counter provided by the GPI interface. The tiles are read from and written back to the data set by one-sided RDMA transfers, which allows a complete overlay of network transfers with computation. Because each tile has to read data at the boundary from its neighboring tiles, read and write accesses to the data set have to be orchestrated. The communication patterns to do so are also implemented completely one-sidedly, not introducing any synchronization point or barrier between any two nodes even inter time steps if the ratio of tiles per node is high enough.

Benchmarks on a cluster with 1,040 compute cores demonstrate the advantages of this implementation. Above a ratio of 4.5 tiles per node the average time that is spent by the nodes on all other activities but computing the tiles is limited to 6%. This figure proves basically constant from 8 nodes (32 cores) up to 256 nodes (1,024 cores). A second and a third benchmark demonstrate the behavior on an inhomogeneous machine simulated by throttling the performance of a single node to 54% and 27% respectively. Above 9 or 18 tiles per node respectively, the communication overhead stays at the 6% level even on these inhomogeneous clusters.

A fourth benchmark demonstrates excellent weak scaling capabilities of our parallelization approach ranging from 8 cores to 1,024 cores. Further good strong scaling is demonstrated over one order of magnitude in number of cores. These results prove the suitability of the GPI programming model for the parallelization of numerical problems in the class of Finite Differences Stencil algorithms.

Acknowledgements We like to thank Joachim Weickert, University of Saarbrücken for providing a single threaded reference implementation of the PSPro Edge- and Coherence-Enhancing Anisotropic Diffusion Filter.

References

1. Bruhn, A., Jakob, T., Fischer, M., Kohlberger, T., Weickert, J., Brüning, U., Schnörr, C.: High performance cluster computing with 3-D nonlinear diffusion filters. Real Time Imag. **10**(1), 41–51 (2004). doi: 10.1016/j.rti.2003.12.002
2. Machado, R., Lojewski, C.: The Fraunhofer virtual machine: a communication library and runtime system based on the RDMA model. Comput. Sci. Res. Dev. **23**(3–4), 125–132 (2009). doi: 10.1007/s00450-009-0088-2
3. Weickert, J.: Theoretical foundations of anisotropic diffusion in image processing. Computing, Suppl **11**, 221–236 (1996)
4. Weickert, J.: Anisotropic diffusion in image processing. Teubner (1998)
5. Weickert, J.: Coherence-enhancing difusion filtering. Internationals Journal of Computer Vision **31**(2/3), 111–127 (1999)

eeClust: Energy-Efficient Cluster Computing

Timo Minartz, Daniel Molka, Michael Knobloch, Stephan Krempel, Thomas Ludwig, Wolfgang E. Nagel, Bernd Mohr, and Hugo Falter

Abstract Energy consumption has become a major topic in high performance computing in the last years. This is first due to high operational costs for large-scale machines, which almost amount to the acquisition costs of the whole installation. A second factor is the high carbon footprint of HPC-centers, which should be reduced for environmental reasons. We present the eeClust project, which aims at the reduction of energy consumption of applications running on a cluster, causing as little performance degradation as possible. We outline the concept of the project, present the tools involved in analyzing the energy consumption of the application as well as managing hardware power states. Further, we present first results and the ongoing work in the project.

1 Introduction

In the last years we have seen a strong increase in energy consumption and related costs in high performance computing. These costs are already in the range of acquisition costs of the whole installation. The project – called Energy-Efficient

T. Minartz (✉) · T. Ludwig
Department of Informatics, University of Hamburg 22527 Hamburg, Germany
e-mail: minartz@informatik.uni-hamburg.de; ludwig@informatik.uni-hamburg.de

D. Molka · W.E. Nagel
Center for Information Services and High Performance Computing (ZIH), TU Dresden 01062 Dresden, Germany
e-mail: daniel.molka@tu-dresden.de; wolfgang.nagel@tu-dresden.de

M. Knobloch · B. Mohr
Jülich Supercomputing Centre (JSC), Forschungszentrum Jülich GmbH 52425 Jülich, Germany
e-mail: m.knobloch@fz-juelich.de; b.mohr@fz-juelich.de

S. Krempel · H. Falter
ParTec Cluster Competence Center GmbH, München
e-mail: krempel@par-tec.com; falter@par-tec.com

C. Bischof (eds.), *Competence in High Performance Computing 2010*,
DOI 10.1007/978-3-642-24025-6_10, © Springer-Verlag Berlin Heidelberg 2012

Cluster Computing[1] – aims at making high performance computing more efficient with respect to economic and ecological aspects. Project partners are the University of Hamburg (coordinator), Dresden University of Technology (TUD/ZIH), Jülich Supercomputing Centre at Forschungszentrum Jülich GmbH, and the ParTec Cluster Competence Center GmbH. Its basic idea is to determine relationships between the behavior of parallel programs and their impact on the energy consumption of the underlying compute cluster. Based on this, strategies will be developed to reduce the energy consumption with as little impact as possible on program performance. In principle, this goal can be achieved when as many as possible hardware components are put into energy saving modes during periods of inactivity. Modern hardware components and operating systems already use these mechanisms based on simple heuristics. However, because no detailed knowledge about the execution behavior of the applications is available, there is a high potential for wrong decisions. The project will develop enhanced parallel programming analysis software based on the successful Vampir and Scalasca software tools. In addition to measuring and analyzing program behavior, they will be enhanced to record energy-related metrics as well. Based on this new energy efficiency analysis, users can insert energy control calls into their applications which will allow the operating system and the cluster job scheduler to control the cluster hardware in an energy-efficient way. The necessary software components will be developed by Hamburg and ParTec. The research team is convinced that results from this project will be applicable in many parallel computing environments. In addition to new tools and mechanisms for controlling energy consumption in relation to program constructs we will also be able to derive new requirements for future hardware and software in high performance computing.

The rest of the paper is structured as follows: The next section introduces the tools involved in this research project. The following Sect. 3 describes the project plan in detail. Section 4 presents first results and an first analysis of the cluster with the SPECPower Benchmark while Sect. 5 shows the ongoing work. Section 6 concludes the paper.

2 Software Components

For achieving the project goal we are using several existing tools developed by the project partners. The tools become extended to work with each other for analyzing and finally controlling a computer cluster to save energy.

[1] eeClust – http://www.eeclust.de.

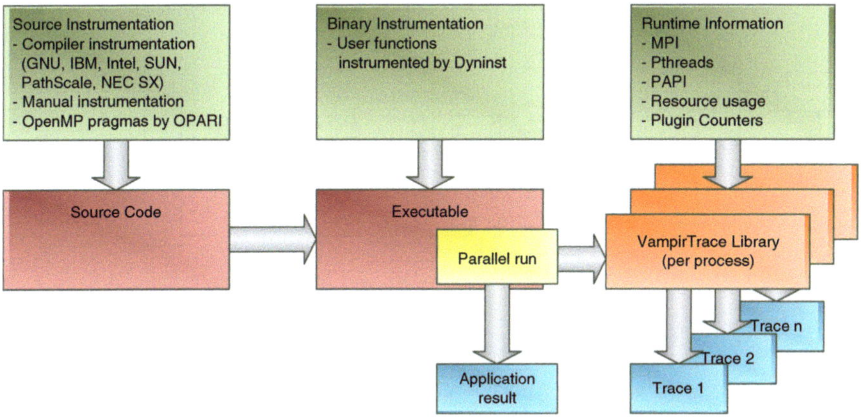

Fig. 1 VampirTrace data sources

2.1 VampirTrace and Vampir

VampirTrace and Vampir are performance analysis and optimization tools developed by the Center for Information Services and High Performance Computing (ZIH). VampirTrace is an open source tool used to instrument applications in order to generate trace files. Depending on the type of instrumentation, the trace files contain events like function entries and exits, MPI messages sent between processes, hardware performance counters, etc. Figure 1 shows the possible data sources.

All events are stored with a timestamp, thus events of different processes can be visualized with the correct timing behavior. Traces are stored in the Open Trace Format (OTF), and the commercial tool Vampir can be used to visualize them (see Fig. 8).

2.2 Scalasca

Scalasca [1], a joint development of Juelich Supercomputing Centre (JSC) and the German Research School for Simulation Sciences (GRS), is an integrated performance analysis toolset to automatically analyze large-scale parallel applications, using the MPI, OpenMP or a hybrid MPI/OpenMP programming model. It has proven scalability up to 294912 processes [2], i.e. the whole BlueGene/P at JSC. The Scalasca workflow is shown in Fig. 2.

The instrumentation of the program and generation of the trace file is very similar to the VampirTrace approach as shown in Fig. 1. In fact, both measurement systems have the same roots and are now united again in the SILC[2] project.

[2]http://www.vi-hps.org/projects/silc/.

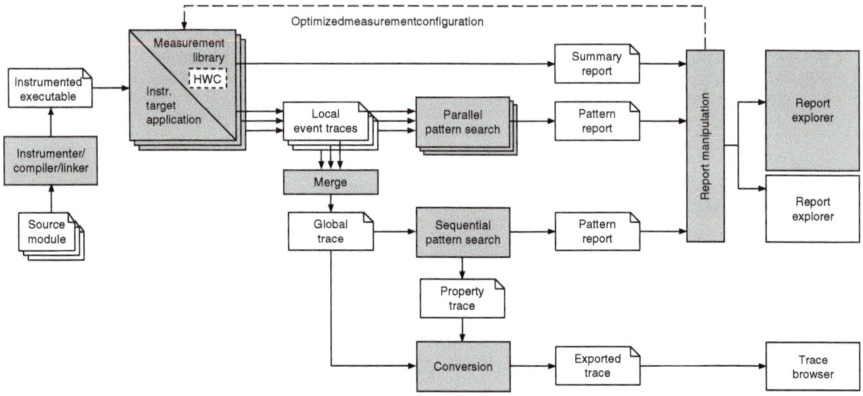

Fig. 2 The Scalasca workflow

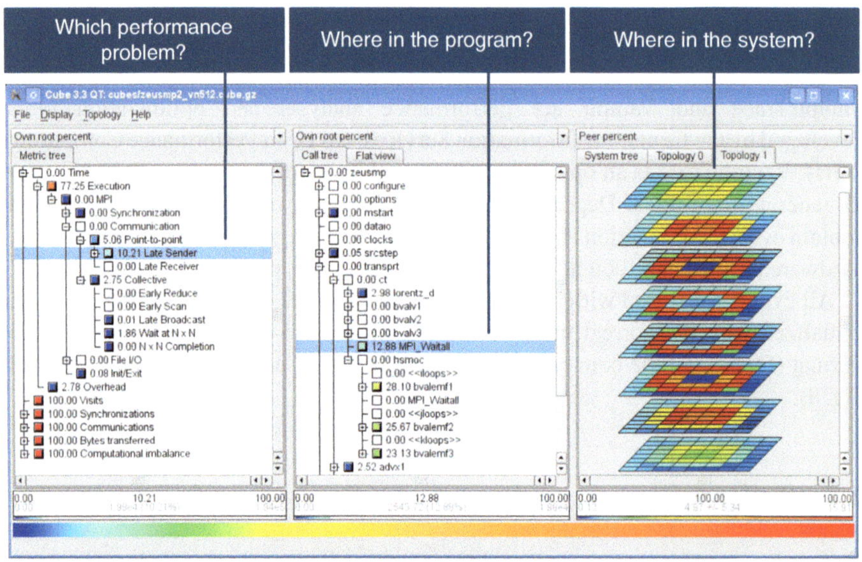

Fig. 3 The CUBE3 Browser

Scalasca includes both a parallel and a sequential trace analyzer, the latter is a remainder from the Scalasca predecessor KOJAK. While the parallel analyzer scales to hundreds of thousands of cores [2], the analysis capabilities of the sequential analyzer are currently still higher, especially for hardware performance counters [3], since it allows a full view on the trace file. However, there is ongoing work to match those capabilities, so that the sequential analyzer will vanish in the long term.

Scalasca supports both profiling and tracing of an application. In either case, the resulting report is displayed in the CUBE3 browser, as shown in Fig. 3.

2.3 GridMonitor

The ParaStation GridMonitor as part of the ParaStationV5 cluster suite is a versatile system monitor for Linux-based compute clusters. A multiplicity of information from different devices and services from a cluster may be read, evaluated and stored. The GridMonitor provides the administrator with various aspects of the available information, from an overall status of all configured clusters to in-depth details of nodes and devices. Data can be grouped with respect to different aspects and are visualized using a web browser. Furthermore, parameters may constantly be monitored and the administrator may be informed, if required.

Data Gathering Process ('Collector')

All available data are retrieved and managed by a so called collector. All data are retrieved using dedicated agents for each device or service. These agents support various protocols, like e.g. SNMP or IPMI. The collector gathers various data from different information sources available within a cluster, describing:

- Compute nodes
- Fileservers and front-end nodes
- Network devices
- Baseboard management controllers
- Storage devices
- Runtime systems

Available parameters not only include operating system values, like system load, network counters or temperatures, but also parameters supplied by runtime systems like batch queuing systems or ParaStation. If available, they may even include information provided by network switches, rack or room environment monitoring devices, uninterruptible power supplies or similar devices. Each parameter is cached within the collector for the configured decay time. Using intelligent caching algorithms, multiple reading of data is avoided. Data can be stored to and retrieved from a database. Thereby, a data history is available, e.g. for plotting diagrams. Each known numerical value can be compared to an upper and lower limit. In case this value under-runs or over-runs those limits, an event will be generated. To constantly monitor these parameters, reading cycles can be defined.

3 Project Plan

Figure 4 describes the project plan in detail. The initial step is to build a power-aware cluster, whose hardware supports multiple operating and idle states. Our cluster consists of five dual socket Intel Nehalem (Xeon X5560, 4 cores + Hyperthreading) and five dual socket AMD Magny-Cours (Opteron 6168, 12 cores) computing

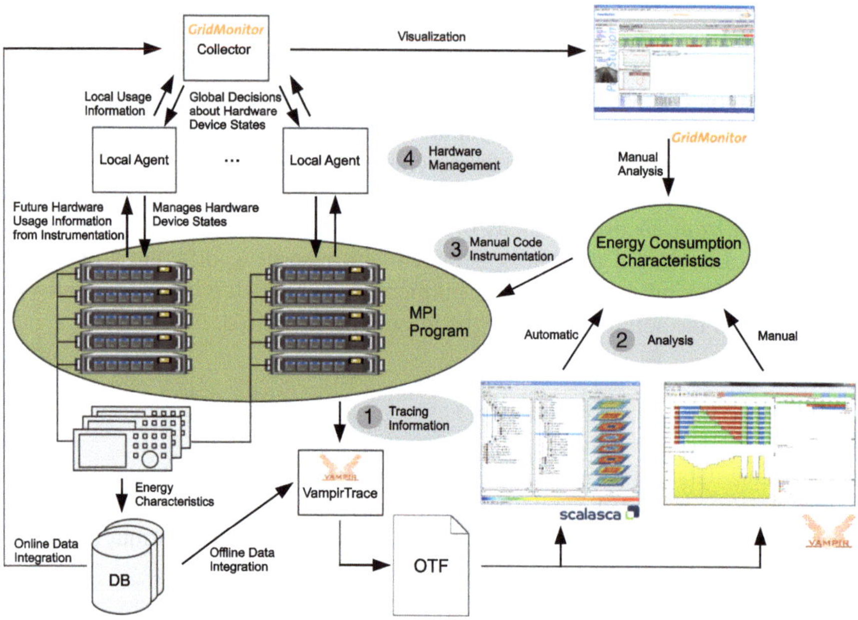

Fig. 4 Project chart of the eeClust Project

nodes. Each processor is DVFS enabled and supports multiple performance states (P-States) and idle states (C-States) [4]. Further, the hard discs and network interface cards of each node support transitions to low power or reduced performance states (D-States). Additionally, the cluster has two I/O subsystems with five disks, one with HDDs and one with SSDs. The computation nodes and the I/O nodes are connected with Gigabit Ethernet. To measure the power consumption of the hardware, each node and the Gigabit switch are connected to ZES LMG450 [5] high precision power meters with an accuracy of about 0.1 %. The power consumption of each node is stored in a database on the head node, to which all power meters are connected via serial ports. The cluster nodes are installed with OpenSuse as operating system, Parastation as cluster management system and VamirTrace, Vampir and Scalasca as tracing and trace analysis tools. The first task is to extend VampirTrace to integrate the energy characteristics from the database, which is done offline after the tracing has been finished. The resulting OTF file then contains the performance characteristics of the parallel program and the resulting power consumption which can be processed in the second step with Vampir and Scalasca. The goal of this analysis is to identify energy consumption characteristics. These characteristics include energy consumption and hardware usage information of specific application code chunks, which will be summarized into a study of energy efficiency code patterns and an energy efficiency benchmark. The third step is the manual instrumentation of the application code to provide future hardware usage information to the agent instances running on each node. The local agent communicates this information to the `Collector`, which can make global decisions about hardware device states based on this information which will be processed the agents. Finally, the hardware device states and the energy characteristics can be visualized using the GridMonitor.

4 Results

This section includes first results of the project. Section 4.1 discusses some runs of the SPECPower [6] benchmark that has been used to examine the impact of certain system settings on the power management efficiency. This benchmark uses typical webserver workloads to generate different levels of utilization on a single node. The following Sect. 4.2 addresses the total power consumption of a cluster that was characterized using SPEC MPI2007 [7]. Section 4.3 describes the extension of VampirTrace that allows to record the power consumption data and add this information to the generated trace file. Section 4.4 introduces microbenchmarks that characterize the dynamic power consumption of arithmetic instructions and data transfers which are major contributors to the dynamic power consumption of processors and main memory.

4.1 *SPECPower*

Figures 5 and 6 show four example runs of the SPECPower benchmark for one of the Intel and AMD nodes respectively.

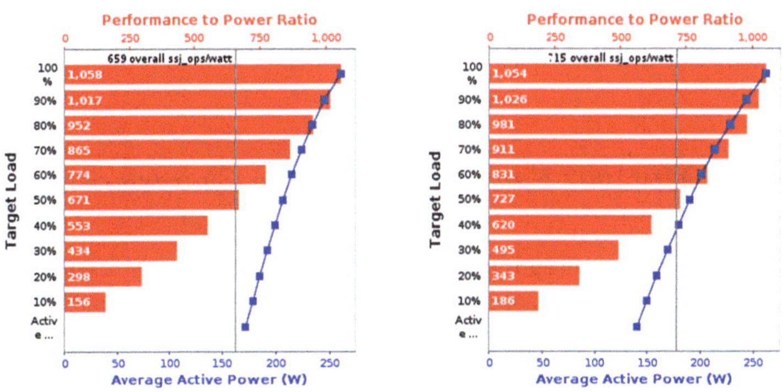

(a) Power saving mechanism deactivated. (b) Power saving mechanism activated.

Fig. 5 SPECPower run on one of the Intel nodes using the *Ondemand Governor* (TurboBoost is disabled)

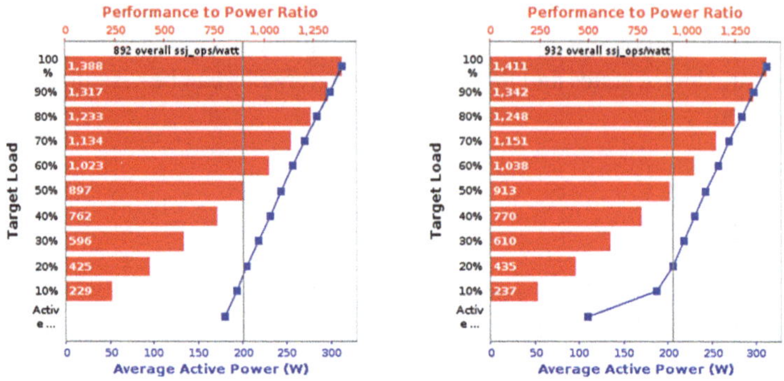

(a) Power saving mechanism deactivated.

(b) Power saving mechanism activated.

Fig. 6 SPECPower run on one of the AMD nodes using the *Ondemand Governor*

For different load levels in 10 % steps the average power consumption in watt and the performance to power ratio in ssj_ops[3] per watt are plotted. When operating under full load, the power consumption with and without power saving mechanism is nearly the same (because the processors can not take advantages of performance and sleep states). Interestingly, the performance to power ratio is also nearly the same under full load – out of this, the performance overhead of energy saving mechanism can be disregarded in terms of efficiency. The more the load is decreased, the energy saving mechanism comes into account and the more the efficiency for the Intel node is increased. The efficiency increase ranges from −0.4 % up to +19.2 %. For the AMD node, the efficiency is varying between +1.0 % and +3.5 %, meaning the processor itself is more efficient (without energy saving mechanism) under low utilization – or, the energy saving mechanisms from AMD are not as effective as the ones from Intel. However, without utilization (active idle) the power consumption of the AMD node increases superlinear, while only linear for the Intel node.

4.2 SPEC MPI2007 Energy Consumption Characteristics

Metering the power consumption of large systems is usually not feasible. One way to estimate the total power consumption of clusters is to extrapolate single node measurements. The Green500 list [8] for example is based on such extrapolations. However, it is doubtful that this method produces accurate results in all cases. The behavior of SPEC MPI2007 [7] on a 32-node IBM iDataplex system has been analyzed [9] to characterize the energy consumption variation of nodes within a

[3] Internal benchmark specific operation.

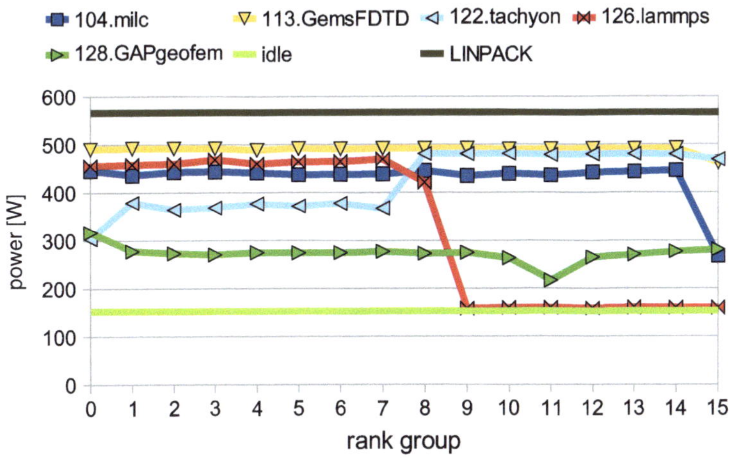

Fig. 7 Comparison of the power consumption of different nodes

clusters. Five out of 13 tested benchmarks show significant variations in node power consumption. These benchmarks are depicted in Fig. 7.

Because of these variations, the energy consumption of the whole cluster cannot be calculated from the measured consumption of a single node multiplied with the number of nodes. However, it is still possible to estimate the consumption of a homogeneous cluster by measuring only a single node or a subset of nodes. Unfortunately, this requires to run the benchmark or application multiple times with shifted assignment of ranks to nodes until each rank was measured exactly once. As the workload distribution between the ranks is usually the same between the runs, each level of utilization can be measured on a single node.

4.3 Recording Energy Consumption with VampirTrace

VampirTrace newly supports a Plugin Interface [10] to add external events into the trace files. This interface has been used to incorporate the power consumption. Figure 8 shows the power consumption of a MPI application during idle (sleep) and compute phases as well as during a MPI barrier. The power consumption is stored in a database during the program execution and added to the trace after the program has finished.

4.4 Influence of Arithmetic Instructions and Data Transfers on Energy Consumption

Synthetic microbenchmarks have been used to study the power consumption characteristics of processors. Power consumption is recorded using a ZES LMG450 power

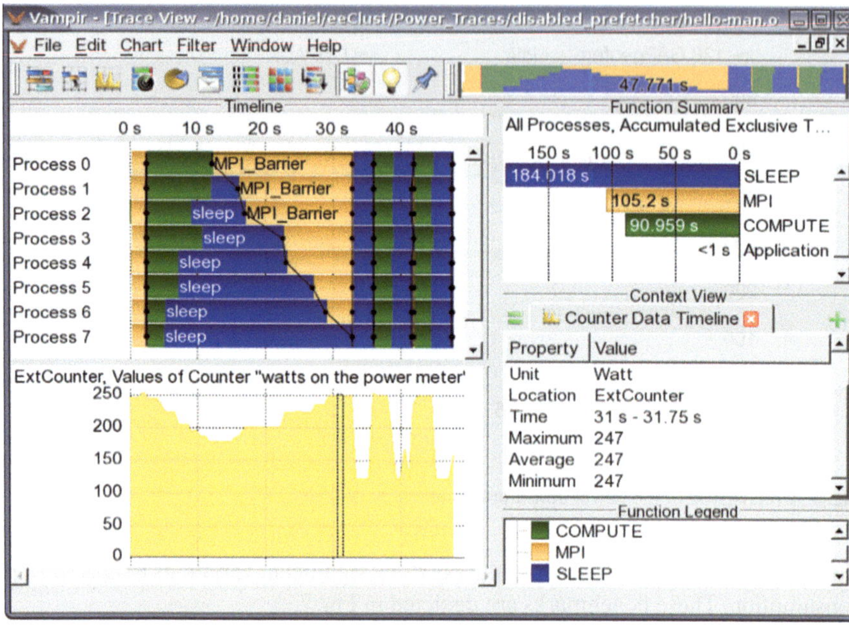

Fig. 8 Power consumption of MPI barrier (OpenMPI 1.3)

meter that measures with a maximal sampling rate of 20 Hz. In order to characterize certain operations with respect to their energy consumption, a constant processor load is required for runtimes that are long enough to be observed with the measurement equipment. The benchmarks used to characterize the power consumption of data transfers are based on open source memory bandwidth benchmarks [11, 12]. These multithreaded benchmarks measure the available bandwidth for each cache level and main memory. They use the `movaps` instruction to stress the memory subsystem without performing arithmetic operations. One appropriate data set size is used for each cache level and main memory. To enable power measurements, the runtime is increased by accessing the whole buffer repeatedly. Additional measurement routines [13] are used to investigate how arithmetic operations contribute to the energy consumption of the CPUs. In these routines the data transfer is replaced with SSE instructions that additionally perform arithmetic operations.

Figure 9 shows the measured power consumption and bandwidths of a two socket Intel Xeon X5670 test system when performing different operations on data in different cache levels. Only the additional power consumed when executing instructions is attributed to the respective operations. An estimated baseline of 220 W is subtracted to account for the power required to keep the system active (i.e. all CPUs in ACPI C0) without performing operations.

Based on our measurements, transfers from the L1 cache consume approximately 64 pJ/Byte. Transfers from the L2 are almost twice as expensive at 121 pJ/Byte. L3 accesses consume 254 pJ/Byte, thus double the required energy again. Transferring

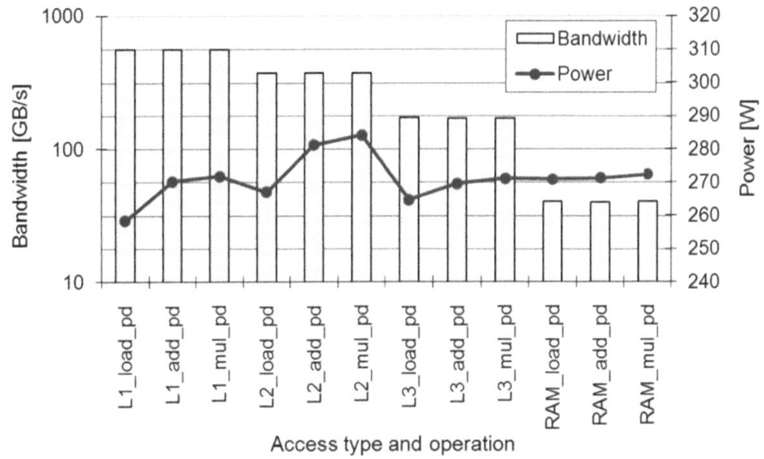

Fig. 9 Power consumption and data throughput on (2 socket Intel Xeon X5670)

Table 1 Energy consumption of arithmetic operations (Intel Xeon X5670)

Workload	operations per instruction	energy per operation
add_pi	2 (64 Bit)	428 pJ/op
mul_pi	2 (64 Bit)	476 pJ/op
add_pd	2 (64 Bit)	319 pJ/op
mul_pd	2 (64 Bit)	387 pJ/op
add_ps	4 (32 Bit)	111 pJ/op
mul_ps	4 (32 Bit)	164 pJ/op

data from main memory consumes 1250 pJ/Byte, i.e five times more than the L3 cache and more than a factor of ten compared to the L1 and L2 caches. Table 1 lists the energy consumption for a selection of arithmetic operations. Interestingly, the Integer operations have the highest energy consumption. Single precision calculations consume less than half the energy of a double precision calculation.

5 Ongoing Work

This section covers the major topics being currently worked on in this project. This include the development of an energy efficiency benchmark, the design and implementation of the node local agent and the integration of Scalasca and GridMonitor.

5.1 Energy Efficiency Benchmark

One of the goals in the eeClust project is to design and implement an energy efficiency benchmark that is able to characterize the power consumption of clusters.

This benchmark should have the ability to selectively stress individual components and thereby investigate the systems ability to effectively limit the power consumption of underutilized units.

The benchmarks used to investigate the power consumption of arithmetic operations and data transfers (see Sect. 4.4) are written in assembler and support only 64 Bit x86 processors. However, an architecture independence implementation is required for the energy efficiency benchmark in order to enable comparisons of wide range of systems. An implementation in a high level language has the additional advantage that different compiler and compiler flags can be compared with respect to the energy efficiency of the generated code. It is currently investigated how the existing kernels can be reimplemented in C in a way that they achieve the same performance like the assembler implementation under the restriction that they only use the generic integer and floating point data types (i.e. no SSE intrinsics). Another limitation of the existing benchmarks is that they only support shared memory systems. However, a MPI based implementation is required for the energy efficiency benchmark in order to analyze clusters. A configurable benchmark generator is currently under development that combines compute kernels with typical communication patterns and I/O activities. Depending on the selected kernels, the communication pattern choices, and I/O parameters the generated benchmarks will utilize the system components differently.

5.2 *Scalasca Integration*

During the project Scalasca will be extended to automatically detect energy saving possibilities in analogy to the detection of wait states in parallel programs. The plan is to automatically detect which components are underutilized and could therefore be put to a lower power state to save energy. With a power model of the machine, i.e. the knowledge of how much energy every single component needs in each power state, a good approximation of the saving possibilities can be calculated.

5.3 *Local Agent*

Most hardware energy management features require superuser privileges in order to change the power state of the component, so applications on a cluster usually are not able to invoke them. Thus, we develop a node-local agent as a daemon, i.e. a background process, outlined in Fig. 10 to provide a uniform interface to hardware management features. The instrumented application then communicates with this daemon, i.e. signals near-term hardware requirements and the daemon invokes energy management features accordingly.

If no application is running on a node, the daemon automatically invokes all power saving features, resulting in a reduced power consumption when a node

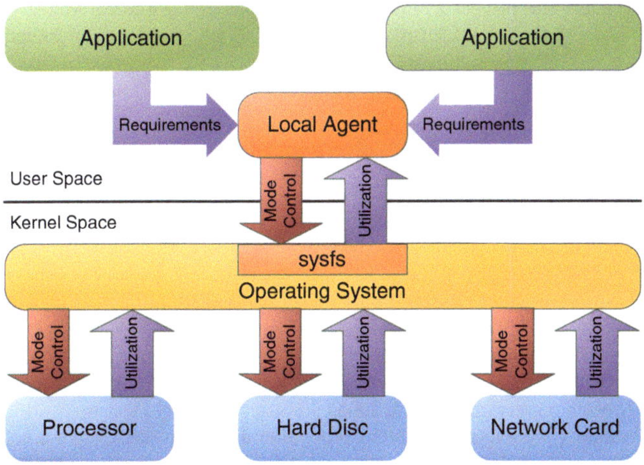

Fig. 10 Scheme of the local agent

is idle. These power saving features do not only include processor power saving features, further the power saving features of the disk and the network card are accessible.

5.4 GridMonitor Integration

The ParaStation GridMonitor is already able to collect and visualize several performance data in a computer cluster. It can also monitor them and trigger events on value over- or under-run. During the project, it shall not only be extended to collect and present energy consumption data, but it will become the central online control component, too. Based on the data collected from the hardware and the instrumented parallel program it will decide which subsystems can be switched into lower power modes and how long they can stay there. The decisions are made based on the results of our research.

6 Conclusion

In this paper we presented the eeClust project in detail and gave an overview of the tools involved in the project. We further showed first results and presented the work currently done in the project.

Energy-efficiency analysis and fine-grained component managing is still basic research and lots of challenges still have to be solved. However, we are convinced that the tools and management mechanism that are developed or extended in this project are able to help cluster operators to achieve significant energy savings.

Acknowledgements This project is funded by the BMBF (German Federal Ministry of Education and Research) under grant 01|H08008E within the call: "HPC-Software für skalierbare Parallelrechner".

References

1. Geimer, M., Wolf, F., Wylie, B.J.N., Abraham, E., Becker, D., Mohr, B.: Concurrency and Computation: Practice and Experience **22**(6), 277 (2010)
2. Wylie, B.J.N., Böhme, D., Mohr, B., Szebenyi, Z., Wolf, F.: In: Proceedings of 24th International Parallel & Distributed Processing Symposium and Workshops (IPDPS, Atlanta, GA, USA) (IEEE Computer Society, 2010)
3. Wylie, B.J.N., Mohr, B., Wolf, F.: Holistic Hardware Counter Performance Analysis of Parallel Programs (Forschungszentrum Jülich, 2005)
4. Hewlett-Packard Corporation and Intel Corporation and Microsoft Corporation and Phoenix Technologies Ltd. and Toshiba Corporation. Advances Configuration and Power Interface Specification (2005)
5. ZES ZIMMER Electronic Systems GmbH. Precision Power Analyzer. URL http://www.zes.com/english/products/index.html/
6. Lange, K.D.: Computer **42**, 95 (2009); DOI http://doi.ieeecomputersociety.org/10.1109/MC.2009.84
7. Müller, M.S., van Waveren, M., Lieberman, R., Whitney, B., Saito, H., Kumaran, K., Baron, J., Brantley, W.C., Parrott, C., Elken, T., Feng, H., Ponder, C.: Concurrency and Computation: Practice and Experience **22**(2), 191 (2010)
8. Feng, W., Cameron, K.: Computer **40**(12), 50 (2007)
9. Hackenberg, D., Schöne, R., Molka, D., Müller, M., Knüpfer, A.: Computer Science - Research and Development **25**, 155 (2010); URL http://dx.doi.org/10.1007/s00450-010-0118-0
10. Schöne, R., Tschüter, R., Hackenberg, D., Ilsche, T.: The VampirTrace Plugin Counter Interface: Introduction and Examples (2010)
11. Molka, D., Hackenberg, D., Schöne, R., Müller, M.S.: In: PACT '09: Proceedings of the 2009 18th International Conference on Parallel Architectures and Compilation Techniques pp. 261–270 IEEE Computer Society, Washington, DC, USA (2009)
12. Hackenberg, D., Molka, D., Nagel, W.E.: In: MICRO 42: Proceedings of the 42nd Annual IEEE/ACM International Symposium on Microarchitecture pp. 413–422. ACM, New York, NY, USA (2009)
13. Molka, D., Hackenberg, D., Schöne, R., Müller, M.S.: In: Proceedings of the 1st International Green Computing Conference pp. 123–133 IEEE (2010)

An Efficient and Flexible Parallel FFT Implementation Based on FFTW

Michael Pippig

Abstract In this paper we describe a new open source software library called PFFT [12], which was developed for calculating parallel complex to complex FFTs on massively parallel architectures. It combines the flexible user interface and hardware adaptiveness of FFTW [7] with a highly scalable two-dimensional data decomposition. We use a transpose FFT algorithm that consists of one-dimensional FFTs and global data transpositions. For the implementation we utilize the FFTW software library. Therefore we are able to generalize our algorithm straight forward to d-dimensional FFTs, $d \geq 3$, real to complex FFTs, and even completely in place transformations. Further retained FFTW features like the selection of planning effort via flags and a separate communicator handle distinguish PFFT from other public available parallel FFT implementations. Automatic ghost cell creation and support of oversampled FFTs complete the outstanding flexibility of PFFT. Our runtime tests up to 262144 cores of the BlueGene/P supercomputer prove PFFT to be as fast as the well known P3DFFT [11] software package, while the flexibility of FFTW is still preserved.

1 Introduction

The fast Fourier transform (FFT) provides the basis of many algorithms in scientific computing. Hence, a highly scalable implementation for massively parallel systems such as BlueGene/P is desirable. There are two approaches to parallelize multi-dimensional FFTs, first binary exchange algorithms, and second transpose algorithms. An introduction and theoretical comparison can be found in [9]. We concentrate on transpose algorithms, i.e., we perform a sequence of local FFTs and

M. Pippig (✉)
Chemnitz University of Technology, Department of Mathematics, 09107 Chemnitz, Germany
e-mail: michael.pippig@mathematik.tu-chemnitz.de

C. Bischof (eds.), *Competence in High Performance Computing 2010*,
DOI 10.1007/978-3-642-24025-6_11, © Springer-Verlag Berlin Heidelberg 2012

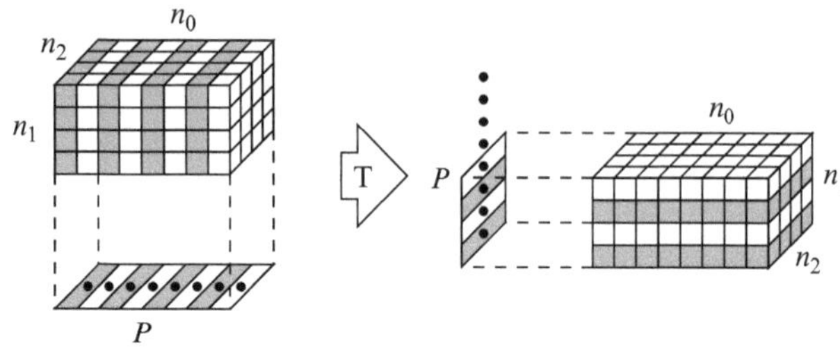

Fig. 1 Distribution of a three-dimensional dataset of size $n_0 \times n_1 \times n_2 = 8 \times 4 \times 4$ on a one-dimensional process grid of size $P = 8$. After the transposition (T) half of the processes remain idle

global data transpositions. For convenience, we consider a three-dimensional input dataset of size $n_0 \times n_1 \times n_2$ with $n_0 \geq n_1 \geq n_2$.

First parallel transpose FFT algorithms were based on a technique called slab decomposition, i.e., the three-dimensional dataset is partitioned along n_0 to distribute it on a given number $P \leq n_0$ of MPI processes. After calculation of the n_0 locally available two-dimensional FFTs of size $n_1 \times n_2$ a data transposition is performed that corresponds to a call of MPI Alltoall and repartitions the three-dimensional dataset along n_1. Finally, $n_1 n_2$ one-dimensional FFTs of size n_0 are computed. For example, implementations of this algorithm are included in the IBM PESSL library [5], the Intel Math Kernel Library [10], and the FFTW [7] software package, which is appreciated for its portable high-performance FFT implementations and flexible user interface. Unfortunately, all of these FFT libraries lack high scalability on massively parallel systems because slab decomposition limits the number of efficiently usable MPI processes by n_1. Figure 1 shows an illustration of the one-dimensional distributed parallel FFT algorithm and an example of its scalability limitation.

A volumetric domain decomposition was used in [1, 2, 4] to overcome the scalability bottleneck and a software library [3] for power of two FFTs customized to BlueGene/L systems was implemented. The dataset is partitioned along two dimensions and therefore the number of MPI processes can be increased to at most $n_1 n_2$. Figure 2 shows an illustration of the two-dimensional distributed parallel FFT algorithm and its improved scalability in comparison to the example in Fig. 1. Portable implementations based on two-dimensional data decomposition are the FFT package from Sandia National Laboratories [13], and the three-dimensional FFT library called P3DFFT [11]. While the FFT algorithm from Sandia National Laboratories is less restrictive on data distributions, runtime tests proved P3DFFT to perform better in most cases. It naturally turns up to ask for a public available parallel FFT library that unifies the flexibility of FFTW and the highly scalable two-dimensional data decomposition.

This paper is divided into the following parts. First we give the necessary definitions and assumptions, which will endure through the whole paper. Next we

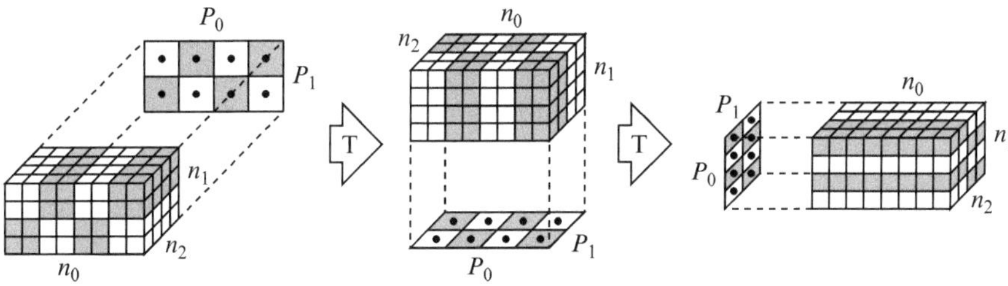

Fig. 2 Distribution of a three-dimensional dataset of size $n_0 \times n_1 \times n_2 = 8 \times 4 \times 4$ on a two-dimensional process grid of size $P_0 \times P_1 = 4 \times 2$. None of the processes remains idle in any calculation step

show the key ideas to implement a parallel FFT algorithm totally based on the FFTW software library. Since P3DFFT is a well known and public available software package for parallel FFTs, we compare features of P3DFFT and our new PFFT package in Sect. 4 and their runtimes in Sect. 5. Finally, we summarize the most important results in Sect. 6.

2 Definitions and Assumptions

Consider a three-dimensional dataset of $n_0 \times n_1 \times n_2$ complex numbers $g_{k_0 k_1 k_2} \in \mathbb{C}$, $k_s = 0, \ldots, n_s - 1$ for all $s = 0, 1, 2$. We write the three-dimensional forward discrete Fourier transform (DFT) as

$$\hat{g}_{l_0 l_1 l_2} := \sum_{k_0=0}^{n_0-1} \sum_{k_1=0}^{n_1-1} \sum_{k_2=0}^{n_2-1} g_{k_0 k_1 k_2} \exp\left(-2\pi \mathrm{i} \left(\frac{l_2 k_2}{n_2} + \frac{l_1 k_1}{n_1} + \frac{l_0 k_0}{n_0}\right)\right) \in \mathbb{C},$$

where $l_s = 0, \ldots, n_s - 1$ for all $s = 0, 1, 2$. It is well known that a multi-dimensional DFT can be calculated efficiently by successive one-dimensional fast Fourier transforms (FFTs).

We assume further the dataset to be mapped onto a two-dimensional process grid of size $P_0 \times P_1$ such that every process owns a block of $n_0/P_0 \times n_1/P_1 \times n_2$ complex numbers. For convenience, we claim n_s to be divisible by P_r for all $s = 0, 1, 2$ and $r = 0, 1$. In order to make the following algorithms more flexible, we can easily overcome these requirements. Depending on the context we interpret the notation n_s/P_r either as a simple division or as a partitioning of the dataset along dimension n_s on P_r processes in equal blocks of size n_s/P_r, for all $s = 0, 1, 2$ and $r = 0, 1$. This notation allows us to compactly represent the main characteristics of data distribution, namely the transposition and the partitioning of the dataset, with diagrams. For example, we interpret the notation $n_2/P_1 \times n_0/P_0 \times n_1$, as transposed dataset of size $n_0 \times n_1 \times n_2$ that is distributed on P_0 processes along the first dimension and on P_1 processes along the last dimension. We assume such multi-dimensional datasets to be stored in C typical row major order, i.e., the last dimension lies consecutively in memory. Therefore, partitioning a multi-dimensional dataset can be done most efficiently along the first dimension.

3 The Parallel Three-Dimensional FFT Algorithm

Our parallel three-dimensional FFT implementation is based on one-dimensional FFTs and two-dimensional data transpositions from FFTW. Therefore, we give a brief overview of the algorithms supported by the FFTW software library and show how to use them to build a parallel three-dimensional FFT.

3.1 One-Dimensional Serial FFT Supported by FFTW

Assume a three-dimensional dataset of $N_0 \times N_1 \times N_2$ complex numbers. The FFTW software library includes algorithms to Fourier transform single dimensions of a multi-dimensional dataset. We sketch those algorithms by the following diagrams

$$\begin{aligned} &N_0 \times N_1 \times N_2 \overset{\mathsf{FFT0}}{\rightarrow} \hat{N}_0 \times N_1 \times N_2, \quad N_0 \times N_1 \times N_2 \overset{\mathsf{FFT2}}{\rightarrow} N_0 \times N_1 \times \hat{N}_2, \\ &N_0 \times N_1 \times N_2 \overset{\mathsf{FFT1}}{\rightarrow} N_0 \times \hat{N}_1 \times N_2, \end{aligned} \tag{1}$$

where FFTX indicates a one-dimensional FFT of dimension X and the hats denote Fourier transformed dimensions. In addition, we are able to combine one-dimensional FFTs with cache oblivious array transpositions [8] by using the most powerful user interface of FFTW. Due to the fact that FFTW can not combine arbitrary in place transpositions with the calculation of one-dimensional FFTs [7], we restrict ourself to the interchange of two successive dimensions. Taking into account that we can substitute two successive dimensions into a single one, we get five different transposition algorithms, which we indicate by the resulting order of dimensions

$$\begin{aligned} &N_0 \times N_1 \times N_2 \underset{\mathsf{012}}{\rightarrow} N_0 \times N_1 \times N_2, \quad (N_0 \times N_1) \times N_2 \underset{\mathsf{201}}{\rightarrow} N_2 \times (N_0 \times N_1), \\ &N_0 \times N_1 \times N_2 \underset{\mathsf{102}}{\rightarrow} N_1 \times N_0 \times N_2, \quad N_0 \times (N_1 \times N_2) \underset{\mathsf{120}}{\rightarrow} (N_1 \times N_2) \times N_0, \\ &N_0 \times N_1 \times N_2 \underset{\mathsf{021}}{\rightarrow} N_0 \times N_2 \times N_1. \end{aligned} \tag{2}$$

Note that all combinations of algorithms from (1) and (2) can be performed in place as well as out of place.

3.2 Two-Dimensional Parallel Transpositions Supported by FFTW

Suppose a two-dimensional dataset of $N_0 \times N_1$ complex numbers is mapped on P processes such that every process holds a block of size $N_0/P \times N_1$. The FFTW3.3alpha1 software library includes a parallel matrix transposition (T) to remap the array into blocks of size $N_1/P \times N_0$. This algorithm is also used for the one-dimensional distributed parallel FFT implementations within FFTW. If we call this library function with the flags FFTW_MPI_TRANSPOSE_OUT and FFTW_MPI_TRANSPOSE_IN, we will be able to handle the following matrix transpositions

$$\begin{aligned} N_0/P \times N_1 &\overset{\mathsf{T}}{\rightarrow} N_1/P \times N_0, \quad N_0/P \times N_1 \underset{\mathsf{OUT}}{\overset{\mathsf{T}}{\rightarrow}} N_0 \times N_1/P, \\ N_1 \times N_0/P &\underset{\mathsf{IN}}{\overset{\mathsf{T}}{\rightarrow}} N_1/P \times N_0, \quad N_1 \times N_0/P \underset{\mathsf{IN,OUT}}{\overset{\mathsf{T}}{\rightarrow}} N_0 \times N_1/P, \end{aligned} \tag{3}$$

where OUT and IN indicate that the corresponding flag was set. There are great advantages of using the parallel transposition algorithms of FFTW instead of direct calls to corresponding MPI functions. FFTW compares different matrix transposition algorithms to get the fastest one. This provides us with portable hardware adaptive communication functions. Furthermore, all transpositions can be performed in place, which is impossible by calls to MPIs standard Alltoall functions and hard to program in an efficient way with point to point communications. In addition, we remain many features of the flexible but easy to use interface of FFTW for our PFFT algorithms. Those features will be explained in more detail in Sect. 4.

3.3 Parallel Three-Dimensional FFT Based on FFTW

Now we suggest a new method for a parallel three-dimensional FFT that can be computed by a careful combination of serial one-dimensional FFTs (1), local data transpositions (2), and global data transpositions (3). Assume a three-dimensional dataset of $n_0 \times n_1 \times n_2$ complex numbers that is distributed on a two-dimensional process grid of size $P_0 \times P_1$. Then, every process holds a local data block of size $n_0/P_0 \times n_1/P_1 \times n_2$. An algorithm to perform a parallel three-dimensional FFT on this dataset is given by the following sequence of one-dimensional FFTs and data transpositions

$$\begin{aligned} n_0/P_0 \times n_1/P_1 \times n_2 &\underset{\mathsf{201}}{\overset{\mathsf{FFT2}}{\rightarrow}} (\hat{n}_2 \times n_0/P_0) \times n_1/P_1 \underset{\mathsf{IN}}{\overset{\mathsf{T}}{\rightarrow}} (\hat{n}_2/P_1 \times n_0/P_0) \times n_1 \\ &\underset{\mathsf{201}}{\overset{\mathsf{FFT2}}{\rightarrow}} (\hat{n}_1 \times \hat{n}_2/P_1) \times n_0/P_0 \underset{\mathsf{IN}}{\overset{\mathsf{T}}{\rightarrow}} (\hat{n}_1/P_0 \times \hat{n}_2/P_1) \times n_0 \\ &\underset{\mathsf{102}}{\overset{\mathsf{FFT2}}{\rightarrow}} \hat{n}_2/P_1 \times \hat{n}_1/P_0 \times \hat{n}_0. \end{aligned}$$

The first step can be obtained by a combination of (1) with the substitutions $N_0 = n_0/P_0, N_1 = n_1/P_1, N_2 = n_2$ and (2) with the substitutions $N_0 = n_0/P_0, N_1 = n_1/P_1, N_2 = \hat{n}_2$, while the second step arises from (3) with the substitutions $N_0 = \hat{n}_2 \times n_0/P_0, N_1 = n_1, P = P_1$. Thereby, mapping a two-dimensional block of size $\hat{n}_2 \times n_0/P_0$ on P_1 processes results in blocks of size $\hat{n}_2/P_1 \times n_0/P_0$, because of row major memory order. All remaining steps can be derived by analogous substitutions. Note that we would have to perform at least two further transpositions to return to the initial data layout. Instead, we expect users to work with transposed data in Fourier space and use the same sequence as above on the transposed dataset $\hat{n}_0 \times \hat{n}_2 \times \hat{n}_1$ and transposed process grid $P_1 \times P_0$ to calculate a three-dimensional backward FFT

$$\begin{aligned}\hat{n}_2/P_1 \times \hat{n}_1/P_0 \times \hat{n}_0 &\underset{201}{\overset{\text{FFT2}}{\rightarrow}} (n_0 \times \hat{n}_2/P_1) \times \hat{n}_1/P_0 \underset{\text{IN}}{\overset{\text{T}}{\rightarrow}} (n_0/P_0 \times \hat{n}_2/P_1) \times \hat{n}_1 \\ &\underset{201}{\overset{\text{FFT2}}{\rightarrow}} n_1 \times (n_0/P_0 \times \hat{n}_2/P_1) \underset{\text{IN}}{\overset{\text{T}}{\rightarrow}} n_1/P_1 \times (n_0/P_0 \times \hat{n}_2) \\ &\underset{102}{\overset{\text{FFT2}}{\rightarrow}} n_0/P_0 \times n_1/P_1 \times n_2.\end{aligned}$$

This gives a considerable gain in performance and, finally, returns to the initial data layout. Of course, there are many other ways to combine the algorithms of FFTW to get a three-dimensional FFT. In our implementation the user is free to choose between different combinations and their resulting data layouts.

4 Parallel FFT Software Library

Our open source parallel FFT software library (PFFT) [12] is available at www.tu-chemnitz.de/~mpip. We now compare the features of our PFFT software library to the P3DFFT software library. Both packages are portable open source libraries which include implementations of parallel three-dimensional FFT based on the highly scalable two-dimensional data decomposition. While P3DFFT is restricted to real to complex FFTs and written in Fortran, our software library calculates complex to complex FFTs and is written in C. Similar to the FFTW software package the user interface of PFFT is split into three layers. The basic interface depends only on the essential parameters of PFFT and is intended to provide an easy start. The possibility to calculate PFFT on a user specified communicator and the ability to change the planning flags of FFTW without recompilation of the PFFT library denote remarkable advantages over the interface of P3DFFT. Both libraries offer in place and out of place FFTs. However, P3DFFT uses buffers roughly three times the size of the input array, while PFFT supports in place algorithms that roughly half the memory consumption in comparison to the out of place algorithms.

More sophisticated adjustments to the algorithm are possible with the advanced user interface. Like P3DFFT we also support a ghost cell creation algorithm to duplicate data near the border at neighboring processes after the FFT calculation finished. In addition, we offer the adjoint ghostcell algorithm that sums up all ghostcells of neighboring processes. As another outstanding feature PFFT gives natural support to oversampled FFTs, i.e., the input array is filled up with zeros to reach the size of the larger output array. If we add all zeros before calling a parallel FFT, we will loose a considerable amount of performance because some processes get parts of the input array that are filled with zeros. Our algorithm omits FFTs on vectors filled with zeros and distributes the work equally. Furthermore, the advanced interface can be easily used to calculate d-dimensional FFTs, where d must be larger than 2, with the two-dimensional data decomposition. That is possible because our algorithms can work on tuples of complex numbers. To calculate a d-dimensional FFT, we call our PFFT library on complex tuples of size $n_3 \times \cdots \times n_{d-1}$ and add one call of a $(d-3)$-dimensional FFT for the last dimensions.

Finally, the guru interface is the right choice for users who want to change the combination of serial FFTs and global data transpositions from Sect. 3.3.

5 Runtime Measurements

In this section we compare the strong scaling behavior of P3DFFT and PFFT up to the full BlueGene/P machine in Jülich Research Center. During the Jülich BlueGene/P Scaling Workshop 2010 we were able to run FFTs of size 513^3 and 1024^3 on up to 64 of the available 72 racks, i.e., 262144 cores. Since P3DFFT only supports real to complex FFTs we applied P3DFFT to the real and imaginary part of a complex input array to get comparable times to the complex to complex FFTs of the PFFT package. The test runs consisted of 10 alternately calculations of forward and backward FFTs. Since these two transforms are inverse except for a constant factor, it is easy to check the results after each run. The average wall clock time as well as the average speedup of one forward and backward transformation can be seen in Fig. 3 for FFT of size 512^3 and in Fig. 4 for FFT of size 1024^3.

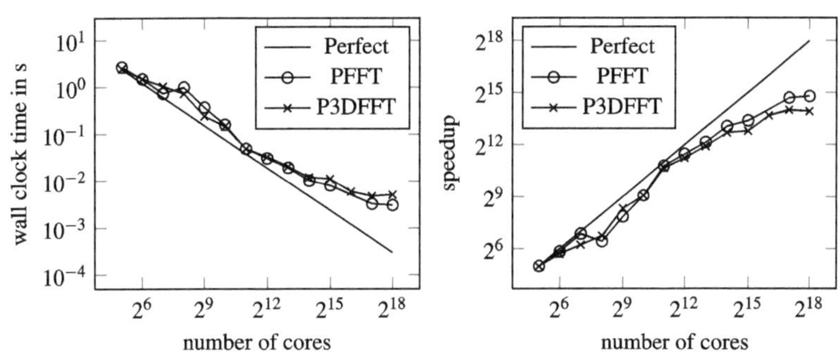

Fig. 3 Runtime measurements for FFT of size 512^3 up to 262144 cores

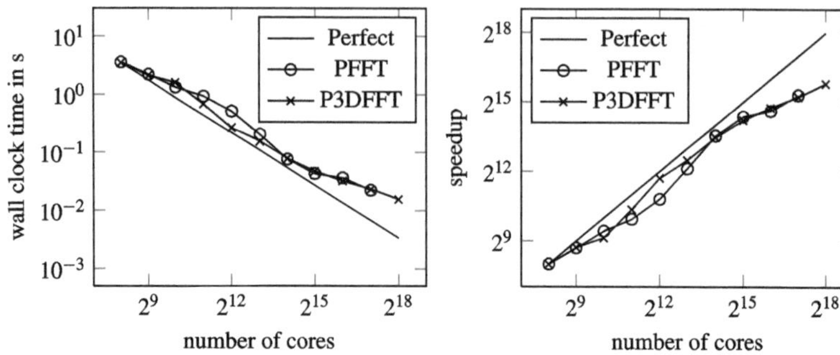

Fig. 4 Runtime measurements for FFT of size 1024^3 up to 262144 cores

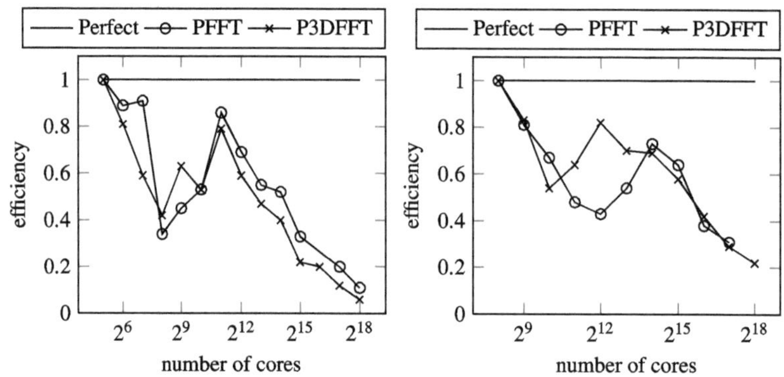

Fig. 5 Efficiency for FFT of size 512^3 (*left*) and 1024^3 (*right*) up to 262144 cores

Memory restrictions force P3DFFT to utilize at least 32 cores on BlueGene/P to calculate a FFT of size 512^3 and 256 cores to perform a FFT of size 1024. Therefore, we chose the associated wall clock times as references for speedup and efficiency calculations. Note that PFFT can perform these FFTs on half the cores because of less memory consumption. However, we only recorded times on core counts which both algorithms were able to utilize to get comparable results.

Unfortunately, the PFFT test run of size 1024^3 on 64 racks died with a RAS event. Nevertheless, our measurements show that the scaling behavior of PFFT and P3DFFT are quiet similar. Therefore, we expect roughly the same runtime for PFFT of size 1024^3 on 64 racks as we observed for P3DFFT.

Note that 262144 is the maximum number of cores we can efficiently utilize for a FFT of size 512^3. This also means that every core calculates only one local FFT of size 512 in each of the three calculation steps. Therefore, the communication takes the largest part of the runtime. The growing communication ratio for increasing core counts also explains the FFT typical decrease of efficiency seen in Fig. 5.

Our flexible PFFT software library can also be used to calculate d-dimensional FFTs, $d \geq 3$. For example, we analyzed the scalability of a four-dimensional FFT

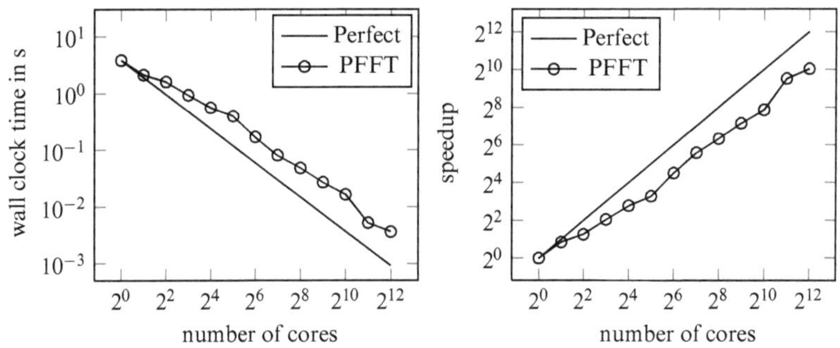

Fig. 6 Runtime measurements for FFT of size 64^4 up to 4096 cores

Fig. 7 Efficiency for FFT of size 64^4 up to 4096 cores

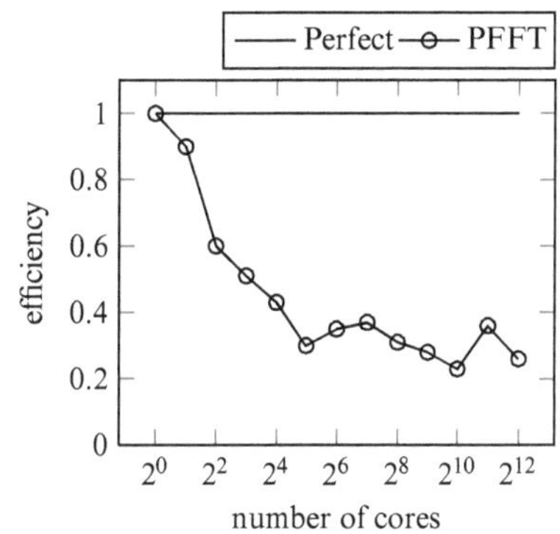

of size 64^4. Since our algorithm uses the two-dimensional data decomposition, we are able to efficiently utilize up to 4096 cores. Note that FFT algorithms based on the one-dimensional decomposition are limited by 64 processes. The runtime measurements given in Figs. 6 and 7 again show the high performance of our PFFT algorithms.

6 Concluding Remarks

We developed an efficient algorithm to compute d-dimensional FFTs ($d \geq 3$) in parallel on a two-dimensional process grid. As well for one-dimensional FFTs as for MPI based communication we exploited highly optimized algorithms of the FFTW software library [6]. Therefore, the interface of our PFFT software library [12] could be easily derived from the flexible user interface of FFTW. This includes the support of parallel in place FFTs without release of high performance. Our runtime tests up to 262144 cores of the BlueGene/P supercomputer prove PFFT to be as fast as the well known P3DFFT software package [11], while the flexibility of FFTW is still preserved. To our knowledge, no public available parallel FFT library has been tested to such great core counts by now. These measurements alleviate the decision-making process, whether a parallel FFT should be used to exploit the full BlueGene/P system.

Acknowledgements This work was supported by the BMBF grant 01IH08001B. We are grateful to the Jülich Supercomputing Center for providing the computational resources on Jülich BlueGene/P (JuGene) and Jülich Research on Petaflop Architectures (JuRoPA).

References

1. Eleftheriou, M., Fitch, B.G., Rayshubskiy, A., Ward, T.J.C., Germain, R.S.: Performance measurements of the 3d FFT on the Blue Gene/L supercomputer. In: Cunha, J.C., Medeiros, P.D. (eds.) Euro-Par 2005 Parallel Processing, Lecture Notes in Computer Science, vol. 3648, pp. 795–803. Springer (2005)
2. Eleftheriou, M., Fitch, B.G., Rayshubskiy, A., Ward, T.J.C., Germain, R.S.: Scalable framework for 3d FFTs on the Blue Gene/L supercomputer: Implementation and early performance measurements. IBM Journal of Research and Development **49**, 457–464 (2005)
3. Eleftheriou, M., Moreira, J.E., Fitch, B.G., Germain, R.S.: Parallel FFT subroutine library. URL http://www.alphaworks.ibm.com/tech/bgl3dfft
4. Eleftheriou, M., Moreira, J.E., Fitch, B.G., Germain, R.S.: A volumetric FFT for BlueGene/L. In: Pinkston, T.M., Prasanna, V.K. (Eds.) HiPC, Lecture Notes in Computer Science, vol. 2913, pp. 194–203. Springer (2003)
5. Filippone, S.: The IBM parallel engineering and scientific subroutine library. In: Dongarra, J., Madsen, K., Wasniewski, J. (Eds.) PARA, Lecture Notes in Computer Science, vol. 1041, pp. 199–206. Springer (1995)
6. Frigo, M., Johnson, S.G.: FFTW, C subroutine library. http://www.fftw.org. URL http://www.fftw.org
7. Frigo, M., Johnson, S.G.: The design and implementation of FFTW3. Proceedings of the IEEE **93**, 216–231 (2005)
8. Frigo, M., Leiserson, C.E., Prokop, H., Ramachandran, S.: Cache-oblivious algorithms. In: Proceedings of 40th Ann. Symp. on Foundations of Comp. Sci. (FOCS), pp. 285–297. IEEE Comput. Soc. (1999)
9. Gupta, A., Kumar, V.: The scalability of FFT on parallel computers. IEEE Transactions on Parallel and Distributed Systems **4**, 922–932 (1993)
10. Intel Corporation: Intel math kernel library. URL http://software.intel.com/en-us/intel-mkl/
11. Pekurovsky, D.: P3DFFT, Parallel FFT subroutine library. URL http://www.sdsc.edu/us/resources/p3dfft
12. Pippig, M.: PFFT, Parallel FFT subroutine library. URL http://www.tu-chemnitz.de/~mpip
13. Plimpton, S.: Parallel FFT subroutine library. URL http://www.sandia.gov/~sjplimp/docs/fft/README.html

ParaSCIP: A Parallel Extension of SCIP

Yuji Shinano, Tobias Achterberg, Timo Berthold, Stefan Heinz, and Thorsten Koch

Abstract *Mixed integer programming (MIP)* has become one of the most important techniques in Operations Research and Discrete Optimization. SCIP (Solving Constraint Integer Programs) is currently one of the fastest non-commercial MIP solvers. It is based on the *branch-and-bound* procedure in which the problem is recursively split into smaller subproblems, thereby creating a so-called *branching tree*. We present ParaSCIP, an extension of SCIP, which realizes a parallelization on a distributed memory computing environment. ParaSCIP uses SCIP solvers as independently running processes to solve subproblems (nodes of the branching tree) locally. This makes the parallelization development independent of the SCIP development. Thus, ParaSCIP directly profits from any algorithmic progress in future versions of SCIP. Using a first implementation of ParaSCIP, we were able to solve two previously unsolved instances from MIPLIB2003, a standard test set library for MIP solvers. For these computations, we used up to 2048 cores of the HLRN II supercomputer.

1 Introduction

Branch-and-bound is a very general and widely used method to solve discrete optimization problems. An important class of problems which can be solved using this method are *mixed integer programs (MIP)*. The challenge of these problems

Y. Shinano · T. Berthold (✉) · S. Heinz · T. Koch
Zuse Institute Berlin, Takustr. 7, 14195 Berlin, Germany
e-mail: shinano@zib.de; berthold@zib.de; heinz@zib.de; koch@zib.de

T. Achterberg
IBM Deutschland GmbH, Ober-Eschbacher Str. 109, 61352 Bad Homburg v.d.H., Germany
e-mail: achterberg@de.ibm.com

C. Bischof (eds.), *Competence in High Performance Computing 2010*,
DOI 10.1007/978-3-642-24025-6_12,

is to find a feasible assignment to a set of decision variables which yields a minimum/maximum value with respect to a given linear objective function. The feasible region for these problems is described by linear inequalities. In addition a subset of the variables are only allowed to take integer values. These problems are NP-hard in general [12].

The well-known idea of branching is to successively subdivide the given problem instance into smaller problems until the individual subproblems (or sub-MIPs) are easy to solve. During the course of the algorithm, a branching tree is generated in which each node represents one of the subproblems. To be able to prune the vast majority of nodes at an early stage, sophisticated mathematical techniques are used. This allows a dramatic reduction of the size of the branching tree. Typically, problems with ten thousand variables and constraints (i.e., approximately 2^{10000} potential solutions) can be solved by investigating a few hundred thousand branch-and-bound nodes.

State-of-the-art MIP solvers such as CPLEX [3], Gurobi [1], or SCIP [8] are based on a branch-and-cut [15] procedure, a mathematically involved variant of branch-and-bound. Parallelizing branch-and-cut algorithms has been proven to be difficult, due to fact that the decisions involved depend on each other [16]. State-of-the-art codes *learn* from the decisions already taken, assuming a sequential ordering. Furthermore, basically all algorithmic improvements presented in the literature aim at reducing the size of the branching tree, thereby making a parallelization less effective and even more difficult. The latter is due to the observation, that they typically increase the need for communication and make the algorithm less predictable. Therefore, a well-designed dynamic load balancing mechanism is an essential part of the parallelizing branch-and-cut algorithms.

Since its introduction in 1992, the MIPLIB [11] has become a standard test set library used to compare the performance of MIP solvers. The MIPLIB contains a collection of difficult real-world instances mostly from industrial applications. Its availability has provided an important stimulus for researchers in this active area. The current version, MIPLIB2003 [4, 9], contains more than thirty unsolved instances when it was originally released. This number could be reduced to six; stalling at this level since 2007. These six instances resisted all attempts of the commercial vendors and the research community to solve them to proven optimality.

Algorithmic improvements for state-of-the-art sequential MIP solvers have been tremendous during the last two decades [10]. For an overview on large scale parallelization of MIP solvers, see [18]. Most of these approaches struggled, however, to catch up with the performance of state-of-the-art commercial and non-commercial sequential MIP solvers when it comes to solving really hard MIP instances of general nature. Many unsolved instances of MIPLIB2003 were first solved using sequential solvers [13].

In the following we describe how we developed a massive parallel distributed memory version of the MIP solver SCIP [5] to harness the power of the HLRN II supercomputer [2] in order to solve two of the remaining open instances of the MIPLIB 2003.

2 SCIP: Solving Constraint Integer Programs

SCIP (Solving Constraint Integer Programs) is a framework for constraint integer programming. Constraint integer programming is an extension of MIP and a special case of the general idea of *constraint programming (CP)*. The goal of SCIP is to combine the advantages and compensate the weaknesses of CP and MIP.

An important point for the efficiency of MIP and CP solving algorithms is the interaction between constraints. SCIP provides two main communication interfaces:

1. propagation of the variables' domains as in CP and
2. the linear programming relaxation as in MIP.

SCIP uses a branch-and-bound scheme to solve constraint integer programs (see Sect. 2.2). The framework is currently one of the fastest non-commercial MIP solvers [14], even so it is suitable for a much richer class of problems. For more details about SCIP we refer to [5, 7, 8].

2.1 *Mixed Integer Programs*

In this paper, we only focus on mixed integer programs (MIPs), which can be defined as follows:

Definition 1 (mixed integer program). Let $\hat{\mathbb{R}} := \mathbb{R} \cup \{\pm\infty\}$. Given a matrix $A \in \mathbb{R}^{m \times n}$, a right-hand-side vector $b \in \mathbb{R}^m$, an objective function vector $c \in \mathbb{R}^n$, a lower and an upper bound vector $l, u \in \hat{\mathbb{R}}^n$ and a subset $I \subseteq N = \{1, \ldots, n\}$, the corresponding *mixed integer program* $\text{MIP} = (A, b, c, l, u, I)$ is to solve

$$\begin{aligned} \min \quad & c^T x \\ \text{s.t.} \quad & Ax \le b \\ & l \le x \le u \\ & x_j \in \mathbb{R} \quad \text{for all } j \in N \setminus I \\ & x_j \in \mathbb{Z} \quad \text{for all } j \in I. \end{aligned}$$

The goal is to find an assignment to the (decision) variables x such that all linear constraints are satisfied and the objective function $c^T x$ is minimized. Note that, the above format is quite general. First, maximization problems can be transformed to minimization problems by multiplying the objective function coefficients by -1. Similarly, "$\ge$" constraints can be multiplied by -1 to obtain "$\le$" constraints. Equations can be replaced by two opposite inequalities.

The *linear programming relaxation* is achieved by removing the integrality conditions. The solution of the relaxation provides a *lower bound* on the optimal solution value.

2.2 *Branch-and-Bound*

One main technique to solve MIPs is the branch-and-bound procedure. The idea of *branching* is to successively subdivide the given problem instance into smaller subproblems until the individual subproblems are easy to solve. The best of all solutions found in the subproblems yields the global optimum. During the course of the algorithm, a *branching tree* is generated in which each node represents one of the subproblems.

The intention of *bounding* is to avoid a complete enumeration of all potential solutions of the initial problem, which usually are exponentially many. For a minimization problem, the main observation is that if a subproblem's lower (dual) bound is greater than the global upper (primal) bound, the subproblem can be pruned. Lower bounds are calculated with the help of the linear programming relaxation, which typically is easy to solve. Upper bounds are obtained by feasible solutions, found, e.g., if the solution of the relaxation is also feasible for the corresponding subproblem.

3 ParaSCIP

In this section, we introduce ParaSCIP, a parallel extension of SCIP. The design goals of ParaSCIP are to exploit SCIP's complete functionality, to keep the interface simple, and to scale to at least 10 000 cores in parallel.

We will focus in this paper on two important features, the dynamic load balancing and the checkpointing mechanism.

3.1 *A Dynamic Load Balancing Mechanism*

In this section we illustrate the workflow of the dynamic load balancing mechanism for ParaSCIP. Workload of a sub-MIP computation strongly depends on two factors. One is the number of branching nodes per solver, which may vary from one to several millions. The other is the computing time of a single branch-and-bound node, which may vary from less than one millisecond to several hours. Therefore, the dynamic load balancing mechanism is a key factor for the parallelization of branch-and-bound algorithms.

3.1.1 Initialization Phase

In the beginning, the LOADCOORDINATOR, which acts as a master process, reads the instance data for a MIP model which we refer to as the *original instance*.

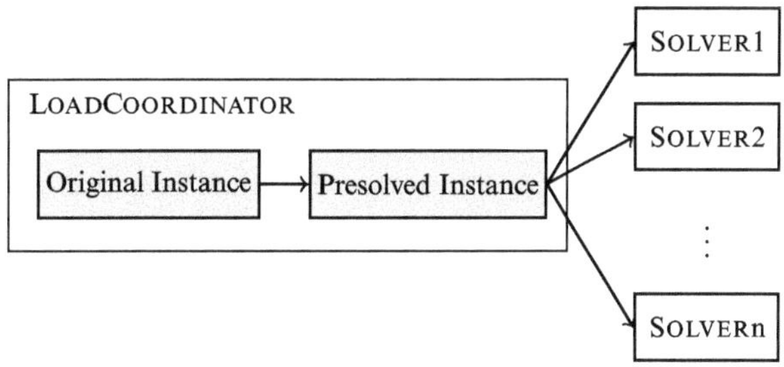

Fig. 1 Initialization phase

This instance is presolved (see Sect. 4.2) directly inside the LOADCOORDINATOR. MIP presolving tries to fix variables and to detect redundancy of certain constraints, for details see [7]. The resulting, typically quite smaller, instance will be called the *presolved instance*. The presolved instance is extracted from the SCIP environment, broadcasted to all available SOLVER processes, and embedded into the (local) SCIP environment of each SOLVER process. This is the only time when the complete instance is transferred. Later, only the differences between a subproblem and the presolved problem will be sent. Figure 1 illustrates this initialization procedure. At the end of this phase all SOLVERs are instantiated with the presolved instance.

3.1.2 Transferring Branch-and-Bound Nodes

After the initialization step, the LOADCOORDINATOR creates the root node of the branch-and-bound tree. Each node transferred through the system acts as the root of a subtree. The information that has to be sent consists only of bound changes for variables between the presolved instance and the subproblem, which gets transferred. For the initial root node there is no difference between the presolved instance and the subproblem.

All nodes, which are transferred to SOLVERs, are kept in the LOADCOORDINATOR with their solver statuses until the corresponding solving process terminates. The SOLVER which receives a new branch-and-bound node instantiates that subproblem using the presolved instance (which was distributed in the initialization phase) and the received bound changes. After that, the SOLVER starts working on the subproblem.

3.1.3 Load Balancing

Load balancing for MIP solving highly depends on the *primal* and *dual bounds*, which are updated during the solving process. The primal bound is given by the value of the best solution that has been found so far during the solving process. If one of the SOLVERs finds an improved solution, this solution is sent to the LOADCOORDINATOR, which distributes the updated primal bound to all other SOLVERs. If a SOLVER receives an improved primal bound, it will immediately apply bounding, hence, prune all nodes in its search tree that cannot contain any better solution anymore.

Periodically, each SOLVER notifies the LOADCOORDINATOR about the number of unexplored nodes in its SCIP environment and the dual bound of its subtree which define the *solver status*. The dual bound is a proven lower bound on the value of the best solution in that subtree. It is derived from the linear programming relaxation (see Sect. 2.1) at the individual nodes. At the same time the SOLVER is notified about the best dual bound value of all nodes in the node pool of the LOADCOORDINATOR, which we will refer to as BESTDUALBOUND. Note that this does not include nodes that are currently processed by any SOLVER.

If a SOLVER is idle and the LOADCOORDINATOR has unprocessed nodes available in the node pool, then the LOADCOORDINATOR sends one of these nodes to the idle SOLVER.

To handle the situation that several solvers become idle at the same time, the LOADCOORDINATOR should always have a sufficient amount of unprocessed nodes left in its node pool. This ensures that the SOLVERs are kept busy throughout the computation.

In order to keep at least p "good" nodes in the LOADCOORDINATOR, we introduce the *collecting mode*, similar to the one introduced in [17]. We call a node *good*, if the dual bound value of its subtree (NODEDUALBOUND) is close to the dual bound value of the complete search tree (GLOBALDUALBOUND).

Consider the case that the LOADCOORDINATOR is not in collecting mode, and it detects that less than p good nodes with

$$\frac{\text{NodeDualBound} - \text{GlobalDualBound}}{\max\{|\text{GlobalDualBound}|, 1.0\}} < \text{Threshold} \tag{1}$$

are available in the node pool, the LOADCOORDINATOR switches to collecting mode and requests selected SOLVERs that have nodes which satisfy (1) to switch also into collecting mode. If the LOADCOORDINATOR is in collecting mode and the number of nodes in its pool that satisfy (1) is larger than $m_p \cdot p$, it requests all collecting mode SOLVERs to stop the collecting mode. Note that p and m_p are parameters which have to be chosen before before a run starts.

If a SOLVER receives the message to switch into collecting mode, it changes the search strategy to either "best estimate value order" or "best bound order" (see [7]). It will then alternately solve nodes and transfer them to the LOADCOORDINATOR. This is done until the SOLVER receives the message to switch out of the collecting mode. If a node of the branch-and-bound tree is selected to be sent to the LOADCOORDINATOR, the corresponding SOLVER collects the bound changes of that node w.r.t. the presolved instance, transfers the differing bounds to the LOAD-COORDINATOR, and prunes the node from the subproblem's branch-and-bound tree.

In the context of parallel branch-and-bound, the process until all SOLVERs are busy is called *ramp-up phase* [18]. In the ramp-up phase, all SOLVERs run in collecting mode. The ramp-up phase continues until the number of nodes in the node pool of the LOADCOORDINATOR is greater than the value p.

The most crucial issue for the load balancing mechanism is to avoid solving useless subproblems. Consider the situation that a SOLVER is solving a subproblem for which the dual bound is already quite large. Then, an improvement in the primal bound will cause all nodes of the subproblem to be pruned.

The SOLVER can detect this situation locally using the best dual bound value of all nodes in the node pool of the LOADCOORDINATOR (BESTDUALBOUND). In this situation, the SOLVER requests another node from the LOADCOORDINATOR while still continuing to solve the current node. After the LOADCOORDINATOR sent a new node to the SOLVER and restored the old solving node in its node pool, the SOLVER stops the solution process and restarts with the new node. The solution of the old node is *delayed*. Note that in case that there is no node available in the LOADCOORDINATOR, the SOLVER keeps continuing to solve the old node.

The termination phase is started after the LOADCOORDINATOR detects that the node pool is empty and all SOLVERs are idle.

3.2 *Checkpointing and Restarting*

Checkpointing mechanisms are a common concept in parallel computing to protect a code against hardware and software failures. The chance that a compute node crashes within a given time frame increases with the number of compute nodes in the parallel system. Further, it is not possible to estimate the computing time for solving a given MIP instance. If the usage of a parallel computing environment is restricted by a certain time limit, we cannot predict reliably whether the computation will be finished within that time window. These two issues make checkpointing and restarting prerequisite functions of a parallel MIP solver.

A natural way for checkpointing would be to save all open nodes of all branch-and-bound trees and the best primal solution found so far. The number of open nodes, however, typically grows very fast for hard problem instances. If checkpointing is performed frequently, this will lead to a huge amount of I/O, slowing down the computation.

Therefore, we decided to save only *primitive* nodes, that is, nodes for which no ancestor nodes are in the LOADCOORDINATOR. This strategy requires much less effort for the I/O system, even in large scale parallel computing environments. For restarting, however, it will take longer to recover the situation from the previous run.

To restart, ParaSCIP reads the nodes saved in the checkpoint file and restore them into the node pool of the LOADCOORDINATOR. After that, the LOADCOORDINATOR distributes these nodes to the SOLVERs ordered by their dual bounds.

In the ramp-up phase, the distributed subproblems are aggressively broken down, because always one of two branched nodes is transferred to the LOADCOORDINATOR. Further, this node will be presolved as it will become the root node of a subproblem. Therefore, the checkpointing and restarting mechanism can be understood as an implicit load balancing mechanism. It detects the hardest part of the branch-and-bound tree and automatically breaks it down to easier subproblems.

4 Solving Open Instances from MIPLIB2003 on HLRN II

In this section, we present computational results for solving two open problem instances, ds and stp3d, from MIPLIB2003 which were conducted on the HLRN II supercomputer [2]. The computations were performed using SCIP 1.2.1.2 with CPLEX 12.1 as underlying linear programming solver.

The best known upper bound for the instance ds and stp3d were 116.59 and 500.736, respectively [13]. The optimal values that we proved are 93.52 for the ds instance (decreased by about 25%) and 493.71965 (decreased by 1%) for stp3d.

4.1 ds and stp3d Instances

The instance ds models a real-world duty scheduling problem of a German public transportation company. In this context, duty scheduling means the assignment of daily shifts of work to bus or tram drivers by means of a schedule. For this particular model, the number of duties, represented by 0-1-variables is 67 732, the number of tasks, represented by linear constraints, is 656. The number of variables set to one in an optimal solution is equal to the cost minimal number of duties to cover all tasks.

stp3d is a Steiner tree packing problem in a three dimensional grid graph. The instance is a "switchbox routing problem" where connections (wires) between various endpoints (terminals) have to been routed in the graph. Each set of endpoints defines a Steiner tree problem which is already NP-hard. In stp3d there are several Steiner trees, which have to be placed at the same time into the graph in a node disjoint way. The objective is to minimize the total length of all networks. Here, already showing feasibility is NP-hard. Consisting of 204 880 variables and 159 488 constraints, stp3d is the largest instance in MIPLIB2003.

4.2 Extended Presolving

Instances that are to be solved on a supercomputer are usually expected to have an enormous running time. Hence, all possibilities to reduce the overall running time in advance should be exploited.

Presolving is an important feature of state-of-the-art MIP solvers that often reduces the overall computation time considerably. The task of presolving is twofold: first, it reduces the size of the model by proving that certain constraints or variables are redundant and can be removed from the problem formulation or fixed to a certain value, respectively, without changing the optimal solution value. Second, it strengthens the LP relaxation of the model by exploiting integrality information, e.g., to tighten the bounds of the variables or to improve coefficients in the constraints. SCIP provides several presolving techniques, some of which are deactivated by default, due to their computational complexity.

Before starting the parallel computation, we applied an *extended preprocessing* on a single machine. Therefore, we used SCIP's "aggressive presolving" settings and afterwards performed strong branching on all problem variables. For 0-1 variables, *strong branching* tentatively fixes a variable to zero (and subsequently to one), and solves the corresponding LP relaxation. If this LP turns out to be infeasible, the variable can be fixed to the opposite value in the original problem. If we could fix at least one of the problem variables, we iterate the process, starting again with aggressive presolving.

In particular for the instance `stp3d`, extended presolving helped to reduce the problem size and thereby the expected computation time. For this instance, the default presolving of SCIP reduces the problem size to 136 500 variables and 97 144 constraints, whereas nine iterations of extended presolving reduced the problem size to 123 637 integer variables and 88 388 constraints.

4.3 HLRN II

HLRN II is a massive parallel supercomputing system which is one of the most powerful computers in Germany and number 64 in the TOP500 list as of November 2010 [6]. From the global system view, HLRN II consists of two identical complexes located at RRZN in Hannover and ZIB in Berlin. Both complexes are coupled by the HLRN link, a dedicated fiber connection for HLRN. In the current stage each complex consists of three parts, the so-called MPP1, MPP2, and the SMP part. We used MPP2 part whose specification is as follows:

- 960 eight-core compute nodes (2 quad-core sockets each for Intel Xeon Gainestown processors (Nehalem EP, X5570) running at 2.93 GHz) with 48 GB memory
- Total peak performance 90 TFlop/s
- Total memory 45 TByte
- 4x DDR Infiniband Dual Rail network for MPI
- 4x DDR Infiniband network for I/O to the global Lustre filesystems

4.4 Computational Results

The instances `ds` and `stp3d` were solved by ParaSCIP on HLRN II using up to 2048 cores for each run and needed to be restarted 16 and 10 times, respectively. Tables 1 and 2 show the status of each job for `ds` and `stp3d`, respectively. HLRN II is not able to swap, hence real memory size is a strict limit. If a memory shortage occurs, all SOLVER processes will terminate immediately and solving has to be restarted.

Tables 1 and 2 show the number of runs needed to solve the instance, the ramp-up times and the elapsed computation times until the final checkpoint of each single run. When a SOLVER finishes a sub-MIP computation, it sends its statistical data to the LOADCOORDINATOR. In case a SOLVER did not finish the first sub-MIP computation by the final checkpoint, its performance will not be taken into

Table 1 Each job status for solving `ds` instance

Job No.	# of cores used	Comp. Time (s)	Ramp-up Time (s)	# of solvers solved at least one sub-MIP	# of nodes to restart	# of nodes solved
1	512	14400.5	249.2	185	0	1328147
2	512	18000.6	247.4	105	1	1766849
3	1024	32401.2	333.1	1019	1	71108615
4	2048	23401.1	535.1	2045	8	137270553
5	2048	14400.9	1335.0	2046	5	88130034
6	2048	18001.1	366.5	2020	89	127816887
7	2048	12600.8	324.8	1997	157	77649716
8	2048	16201.2	333.0	1976	248	102950355
9	2048	14401.0	282.0	1927	311	87778722
10	2048	12600.9	309.0	1930	301	71857862
11	2048	12600.8	299.0	1937	292	74279899
12	2048	14400.9	277.5	1911	219	90106215
13	2048	30673.3	–	57	243	477
14	2048	10800.7	294.3	1917	191	57494144
15	2048	14401.0	300.5	1935	196	87130452
16	2048	41402.7	363.4	2047	256	97785724
17	1024	8820.1	353.2	1023	196	363472

– : stopped before ramp-up

Table 2 Each job status for solving `stp3d` instance

Job No.	# of cores used	Comp. Time (s)	Ramp-up Time (s)	# of solvers solved at least one sub-MIP	# of nodes to restart	# of nodes solved
1	512	41467.4	6165.8	239	0	64545
2	1024	41407.4	2927.4	320	146	199719
3	2048	41403.9	2362.3	1185	592	766133
4	2048	1800.3	–	83	527	100
5	2048	1800.4	–	2	446	2
6	2048	1800.3	–	1	444	1
7	2048	1800.9	–	1	443	1
8	2048	43203.9	2474.6	822	442	806159
9	2048	41403.9	2914.2	1817	626	1555160
10	2048	41405.6	4406.7	1778	229	1538151
11	2048	152912.0	3841.8	2047	429	9858917

– : stopped before ramp-up

consideration for the statistics. The tables show the number of SOLVERs that solved at least one sub-MIP to optimality. Note that the SOLVERs which have been idle until the final checkpoint, e.g. because ramp-up did not finish, have no statistical data. The "# of nodes to restart" column shows the number of nodes used for restarting the computation. This corresponds to the number of nodes that had been saved at the final checkpoint of the previous run. Thus, it is always zero for the first job and greater zero for all subsequent jobs. The final column shows the number of nodes solved at the job. Note that the number of nodes at the checkpoints is very small compared to the number of nodes solved for the computations.

The ramp-up time varies among different runs for the same instance. This is due to the fact, that we changed some parameter settings to better adopt to the individual behavior of an instance. These changes only influence the path the solver takes, not the overall result. It took approximately 86 hours to solve `ds` and approximately 114 hours to solve `stp3d` to proven optimality. Due to the checkpoint system described above, some parts of the tree might be resolved several times and only the final solve will be counted in the statistics.

The summary of statistical data therefore gives an underestimation of the number of branch-and-bound nodes. The number of nodes was 1 174 818 123 for `ds` and 14 788 888 for `stp3d`. All SOLVERs bookkeep the idle time, that is not used for solving any sub-MIP. The idle time ratio of all SOLVERs was about 2.2% for solving `ds` and 3.9% for solving `stp3d`. We plan to conduct a single job computation to solve these instances, in order to provide precise values for the number of branch-and-bound nodes and the idle time ratios.

Figures 2 and 3 show how the primal and dual bound evolved during the course of the solution process. The behavior is typical for MIP instances. The primal bound moves stepwise and reaches the optimal values significantly faster than the dual

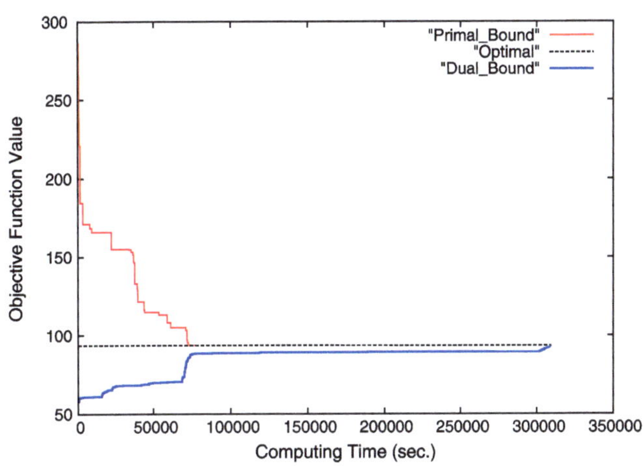

Fig. 2 Bounds evolution for `ds`

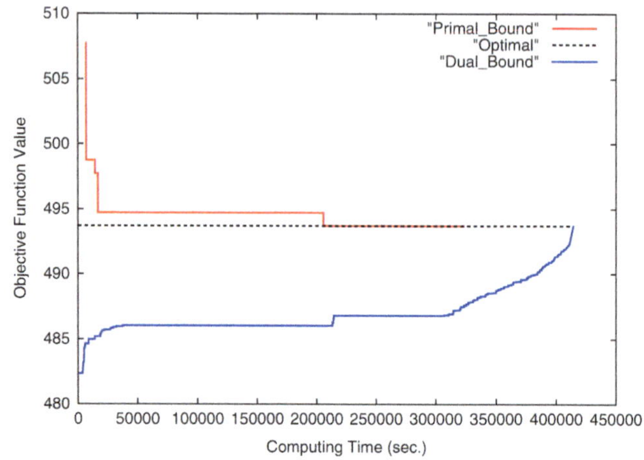

Fig. 3 Bounds evolution for `stp3d`

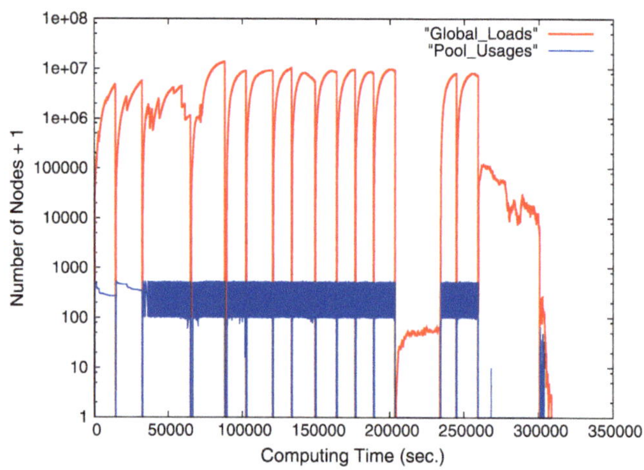

Fig. 4 Workload evolution for `ds`

bound. The dual bound moves smoothly, stays nearly constant for a long time and collapses towards the end of the solution process.

Figures 4 and 5 show how the workloads (that is, the number of branch-and-bound nodes left in all SOLVERS) and the number of nodes in the LOADCOORDINATOR node pool change during the computations. During the whole solution process, nearly all nodes are "good" w.r.t. our definition from Sect. 3.1. ParaSCIP manages very well to keep the node pool of the LOADCOORDINATOR filled. As long as the node pool is not empty, no SOLVER will become idle. Thus, the small idle time ratio is due to the management of the node pool.

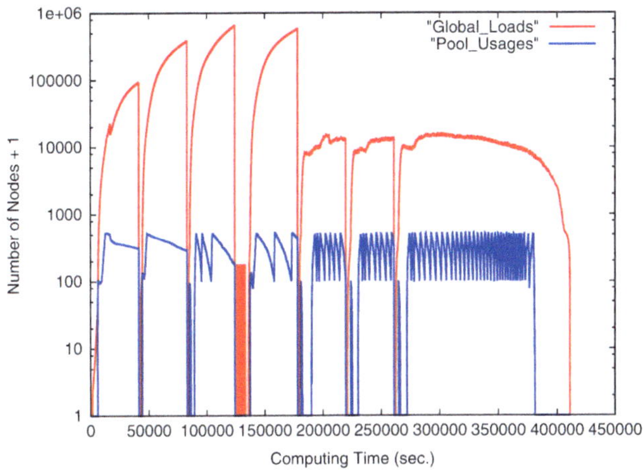

Fig. 5 Workload evolution for `stp3d`

5 Concluding Remarks

We have shown that using our approach, ParaSCIP is able to effectively use the computing power scales of several thousand cores to solve mixed integer programs. Furthermore, this could be done without changing the inner workings of the sophisticated sequential algorithm.

Still, many open questions remain. Using a shared memory version of ParaSCIP, we are planning to estimate how much speed we loose by not using a more fine grained parallelism. When increasing the number of cores, the time spent in the ramp-up and ramp-down phases also increases, hampering scalability. We are currently investigating ways to improve the effectiveness of load balancing during these phases. While `ds` and `stp3d` could be successfully solved, others instances remain, for which it is unclear which amount of computing time is needed to produce an optimal solution with today's MIP solver technology. Trying to judge in advance, whether an instances is suitable for massive parallel solving and predicting remaining running times is an important topic to be investigated in the future.

Acknowledgements Supported by the DFG Research Center MATHEON *Mathematics for key technologies* in Berlin. We are thankful to the HRLN II supercompter stuff, a specially Bernd Kallies and Hinnerk Stüben which gave us support at any time we needed it.

References

1. Gurobi Optimizer. http://www.gurobi.com/
2. HLRN – Norddeutscher Verbund zur Förderung des Hoch- und Höchstleistungsrechnens. http:/ www.hlrn.de/

3. IBM ILOG CPLEX Optimizer. http://www-01.ibm.com/software/integration/optimization/cplex-optimizer/
4. Mixed Integer Problem Library (MIPLIB) 2003. http://miplib.zib.de/
5. SCIP: Solving Constraint Integer Programs. http://scip.zib.de/
6. TOP500 Supercomputer Sites. http://www.top500.org/list/2010/11/100
7. Achterberg, T.: Constraint integer programming. Ph.D. thesis, Technische Universität Berlin (2007)
8. Achterberg, T.: SCIP: Solving constraint integer programs. Mathematical Programming Computation **1**(1), 1–41 (2009)
9. Achterberg, T., Koch, T., Martin, A.: MIPLIB 2003. Operations Research Letters **34**(4), 1–12 (2006)
10. Bixby, R., Rothberg, E.: Progress in computational mixed integer programming – A look back from the other side of the tipping point. Annals of Operations Research **149**(1), 37–41 (2007)
11. Bixby, R.E., Boyd, E.A., Indovina, R.R.: MIPLIB: A test set of mixed integer programming problems. SIAM News **25**, 16 (1992)
12. Karp, R.M.: Reducibility among combinatorial problems. In: R.E. Miller, J.W. Thatcher (eds.) Complexity of Computer Computations, pp. 85–103. Plenum Press, New York, USA (1972)
13. Laundy, R., Perregaard, M., Tavares, G., Tipi, H., Vazacopoulos, A.: Solving hard mixed-integer programming problems with Xpress-MP: A miplib 2003 case study. INFORMS Journal on Computing **21**(2), 304–313 (2009)
14. Mittelmann, H.: Mixed integer linear programming benchmark (serial codes). http://plato.asu.edu/ftp/milpf.html
15. Padberg, M., Rinaldi, G.: A branch-and-cut algorithm for the resolution of large-scale symmetric traveling salesman problems. SIAM Review **33**, 60–100 (1991)
16. Ralphs, T.K., Ladányi, L., Saltzman, M.J.: Parallel branch, cut and price for large-scale discrete optimization. Mathematical Programming Series B **98**(1–3), 253–280 (2003)
17. Shinano, Y., Achterberg, T., t: Fujie: A dynamic load balancing mechanism for new paralex. In: Proceedings of ICPADS 2008, pp. 455–462 (2008)
18. Xu, Y., Ralphs, T.K., Ladányi, L., Saltzmann, M.J.: Computational experience with a software framework for parallel integer programming. INFORMS Journal on Computing **21**(3), 383–397 (2009)

High-Performance Computing Techniques for Coupled Fluid, Structure and Acoustics Simulations

Dörte C. Sternel, Michael Kornhaas, and Michael Schäfer

Abstract A framework for fully coupled numerical simulation of fluid flow, structural deformation, and acoustics is presented. The methodology involves an implicit partitioned approach for fluid-structure interaction, a viscous-acoustic splitting technique for flow acoustics, and corresponding coupling schemes. All components are designed for the use on parallel high-performance computers. Special emphasis is given to the use of geometric multi-grid techniques in order to increase the efficiency of the numerical simulations. Results for several test cases illustrate the capabilities of the approaches considered.

1 Introduction

The immense growth of computational resources enables the simulation of more complex problems. Multi-physics become feasible and concepts for handling the coupling of the simulation of different physical phenomena in different scales of time and space are required. One application of interest is the aeroacoustic noise generation of deformable structures in turbulent flows, as they occur at fans, air conditioning systems, structures of driving cars, etc. Even though the prediction and reduction of noise is of great importance and an important issue during the design process, the physical mechanisms of noise generation in turbulent flows are still not fully understood for many applications – especially when both aeroacoustic and vibro-acoustic effects or fluid induced structural deformations and/or vibrations play a role. Computational fluid dynamics (CFD), in particular time resolved methods like the Direct Numerical Simulation (DNS) and the Large Eddy Simulation (LES) can be a powerful tool for a better understanding of flow physics and the

D.C. Sternel (✉) · M. Kornhaas · M. Schäfer
Institute of Numerical Methods in Mechanical Engineering, Dolivostraße 15, 64293 Darmstadt, Germany
e-mail: sternel@fnb.tu-darmstadt.de

C. Bischof (eds.), *Competence in High Performance Computing 2010*,
DOI 10.1007/978-3-642-24025-6_13, © Springer-Verlag Berlin Heidelberg 2012

mechanisms of aerodynamic noise generation. This is not only because of their capability of predicting fluctuating quantities that can be used as source terms for computational aero-acoustics (CAA) simulations, but also because of their higher accuracy compared to the numerically cheaper Reynolds Averaged Navier-Stokes (RANS) simulations. But for some applications also fluid induced deformations and vibrations – the fluid structure interaction (FSI) – has to be taken into account to be able to correctly predict the physical behavior.

The simulation of the phenomena above, in particular, the necessity of time resolved turbulent scales, requires a large amount of computational resources and the use of highly efficient and parallelized methods is indispensable. In this article we focus on the use of geometric multi-grid techniques in an FSI-framework and the use of its data infra structure for the acceleration of coupled CFD-CAA simulations. More extensive descriptions of the methods, their validation, efficiency investigations, and further acceleration techniques can be found in [5, 8, 17].

2 Fundamental Equations and Model Approaches

The presented approaches consider the coupling of the flow of Newtonian fluids in a range of low Mach numbers (Ma < 0.3) to acoustics and structural movements. For the formulation of the fluid-structure-interaction an implicit partitioned approach is used, for which the fluid domain Ω_F and the structure domain Ω_S are not overlapping. For the acoustics the same domain as for the flow ($\Omega_A = \Omega_F$) is considered.

In this section, the basic equations for the individual physical phenomena are presented.

2.1 Flow

For Newtonian fluids the flow field is described by the following equations:

$$\frac{\partial \rho_F}{\partial t} + \frac{\partial(\rho_F u_i)}{\partial x_i} = 0, \tag{1}$$

$$\frac{\partial(\rho_F u_i)}{\partial t} + \frac{\partial(\rho_F u_i u_j)}{\partial x_i} = \frac{\partial \tau_{Fij}}{\partial x_j} + \rho_F f_i. \tag{2}$$

u_i is the flow velocity vector with respect to Cartesian coordinates x_i, t ist the time, ρ_F is the fluid density and f_i are the external volume forces. τ_{Fij} denotes the Cauchy stress tensor for incompressible Newtonian fluids, which is defined by

$$\tau_{Fij} = \mu_F \left(\frac{\partial u_i}{\partial x_j} + \frac{\partial u_j}{\partial x_i} \right) - p\delta_{ij}, \tag{3}$$

with the pressure p, the dynamic viscosity μ_F and the Kronecker delta δ_{ij}.

For low Mach numbers (Ma < 0.3) the fluid can be handled as incompressible and Eq. (1) reduces to

$$\frac{\partial u_i}{\partial x_i} = 0. \tag{4}$$

Because of the required fine discretization for resolving all scales of turbulent motion, the direct numerical simulation (DNS) of Eqs. (1)–(2) are only feasible for low Reynolds numbers. In the case of higher Reynolds numbers, modeling of turbulence is essential. For all turbulent flows considered, the turbulence is handled by *Large Eddy Simulation* (LES).

2.2 Flow and Aeroacoustics

Generally, the acoustic field can be described by the compressible Navier-Stokes equations (Eqs. (1)–(2)). When an acoustic field superimposes a flow in a low Mach range, it is possible to split the flow variables in an incompressible hydrodynamic part (denoted by ^{inc}) and the acoustic fluctuations (denoted by ^{ac}) [15, 16]:

$$u_i = u_i^{inc} + u_i^{ac} \tag{5}$$

$$p = p^{inc} + p^{ac} \tag{6}$$

$$\rho_F = \rho_F^{inc} + \rho_F^{ac}. \tag{7}$$

Using this "acoustic-viscous-splitting", some transformations, assumptions on homogeneity and isentropy, and neglecting viscous effects in the acoustics, the following equations can be deduced (for a detailed description see e.g. [8]):

$$\frac{\partial \rho_F^{ac}}{\partial t} + \rho_F^{inc} \frac{\partial u_i^{ac}}{\partial x_i} + u_i^{inc} \frac{\partial \rho_F^{ac}}{\partial x_i} = 0 \tag{8}$$

$$\rho_f^{inc} \frac{\partial u_i^{ac}}{\partial t} + \rho_F^{inc} u_j^{inc} \frac{\partial u_i^{ac}}{\partial t} + \frac{\partial p^{ac}}{\partial x_i} = 0 \tag{9}$$

$$\frac{\partial p^{ac}}{\partial t} - \kappa p^{inc} \frac{\partial u_i^{ac}}{\partial x_i} + u_i^{inc} \frac{\partial p^{ac}}{\partial x_i} = -\frac{\partial p^{inc}}{\partial t}. \tag{10}$$

κ denotes the adiabatic exponent. On the left hand side, this system of equations is equivalent to the linearized Euler equations (LEE). The term on the right hand side is the acoustic source, produced by the incompressible flow.

2.3 Structure

While the above equations are in an Eulerian formulation, the structural part is formulated with a Lagrangian view. For the solid part Ω_S, with an assumed constant density ρ_S, the equation of motion is than given by:

$$\rho_S \frac{D^2 d_i}{Dt^2} = \frac{\partial \tau_{S ij}}{\partial x_j} + \rho_S f_i . \tag{11}$$

d_i denotes the displacement vector and $\tau_{S ij}$ is the Cauchy stress tensor. For the considered test cases the St. Venant-Kirchoff material model is employed [18].

3 Numerical Realization

In this section, first the tools – on which the numerical realizations are based – are shortly introduced. Next, the realization of the flow-structure coupling and of the flow-acoustics coupling are described.

3.1 Basic Tools

The flow solver FASTEST [7] constitutes the basis for the fluid-structure-interaction (FSI) approach as well as for the flow-acoustic coupling. The flow solver FASTEST involves a finite volume discretization on a block-structured hexahedral mesh, employing a second-order Taylor-based spatial interpolation [12]. Due to the moving structure, and hence the moving coupling interfaces, the discretization employs an arbitrary Lagrange-Euler formulation. The integration of Equations 1-2 over a volume whose boundaries vary with time, is performed in accordance with the space conservation law [2]. The code works with a SIP-solver in a full multi-grid environment and with implicit time discretization schemes of second-order accuracy (Crank-Nicolson or BDF2). The code is fully parallelized by means of a block structured grid partitioning approach with automatic load balancing. For the simulation of turbulence, FASTEST includes the Smagorinsky model with the dynamic approach of Germano et al. [4] for the *Large Eddy Simulation* (LES). Details of implementation are described in [3].

For the simulation of structural movements, the program FEAP [18] is employed. FEAP is a finite element code in which standard element types are available.

The FSI is realized via an implicit partitioned approach. For this, FASTEST is coupled via the coupling interface MpCCI [13] to FEAP. MpCCI controls the data communication and also provides a conservative interpolation from the spatial fluid discretization to the solid discretization at the coupling interface.

The computation of the acoustic propagation is done by integrating an acoustics solver into the program structure of FASTEST. Details are described below.

3.2 Fluid-Structure Interaction

A schematic view of the coupling scheme is given in Fig. 1. Each time step consists of a nested iteration procedure for the force-displacement coupling. The fluid solver

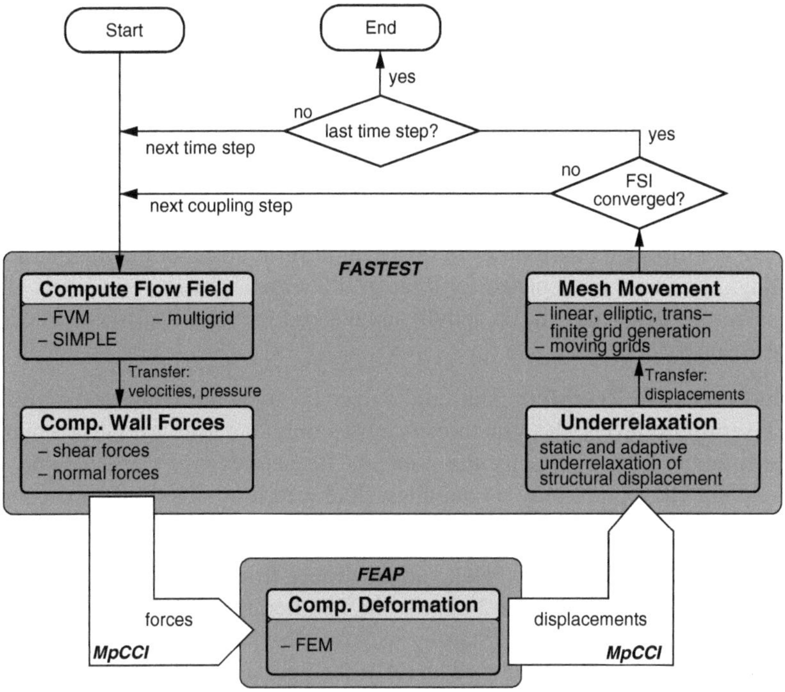

Fig. 1 Flow chart of coupled solution procedure; one time step

FASTEST starts first and calculates the flow field, using the initial geometry. From this, the friction and pressure forces, are computed and passed to the structural code as boundary conditions. The structural solver FEAP computes the deformations with which the fluid mesh is then modified. The calculated displacements at the coupling faces are given via MpCCI to FASTEST, before the flow solver is started again. The first step in the flow solver is the mesh movement, for which specially designed algebraic and elliptic techniques are employed (see [20]).

The actually computed displacements d_i^{act} are linearly weighted with the values d_i^{old} from the preceding iteration to give the new displacements d_i^{new}:

$$d_i^{\mathrm{new}} = \alpha_{\mathrm{FSI}} d_i^{\mathrm{act}} + (1 - \alpha_{\mathrm{FSI}}) d_i^{\mathrm{old}}, \tag{12}$$

where $0 < \alpha_{\mathrm{FSI}} \leq 1$. An acceleration of the computations can be achieved by applying an adaptive determination of α_{FSI} and a predictor for the displacements (see [14]).

The nested iteration procedure continues until it reaches the convergence criterion $\varepsilon_{\mathrm{FSI}}$, defined as follows:

$$\frac{\left\| d^{m-1} - d^{m} \right\|_{\infty}}{\left\| d^{m} \right\|_{\infty}} < \varepsilon_{\mathrm{FSI}}, \tag{13}$$

where m is the number of the FSI iteration and $\|\cdot\|_{\infty}$ denotes the uniform norm.

Geometric multi-grid methods [1] are well established for flow simulations. To use the multi-grid method for fluids with moving boundaries, additionally a flux correction term must be considered in the restriction and prolongation steps [21].

There are different possibilities to realize multi-grid ideas for the fluid-structure coupling. The ideas are depicted for three grid levels. The multi-grid levels of the flow solver are denoted by h, 2h and 4h and the grid levels of the structure code by *h-A*, *h-B* and *h-C*.

Single-Level (SL) Transfer: The fluid solver involves the multi-grid algorithm with several grid levels, the structure code uses only one grid level. The coupling to the structure code is exactly the same, as for single- grid computations, i.e., the coupling takes place only on the finest fluid grid level. Thus, the procedure is represented by the flow chart in Fig. 1.

Multi-Level (ML) Transfer: This variant differs from the single-level transfer in the way of coupling. Figure 2 shows one FSI-Iteration, which consists of one V-cycle, with this modification. Starting with the flow solver on the finest grid, the actual mesh is propagated to all grid levels of the flow solver. After a given number of iterations on grid level h, the wall forces are computed and passed via MpCCI to the structure code. Here, the resulting deformations are computed on grid level *h-A*, stored on grid level *h-A* and given back via MpCCI to the flow solver in which the mesh deformations are computed and stored for this grid level. Now the standard restriction of velocities and residuals to the grid level 2h is done. On the second grid level, the procedure starts again, by doing some iterations and passing the wall forces to the structure solver on grid level *h-B*. The deformations are calculated and stored there, too, and the deformations are given back to the flow solver. The updated mesh is stored for grid level 2h and so on, down to the coarsest grid level.

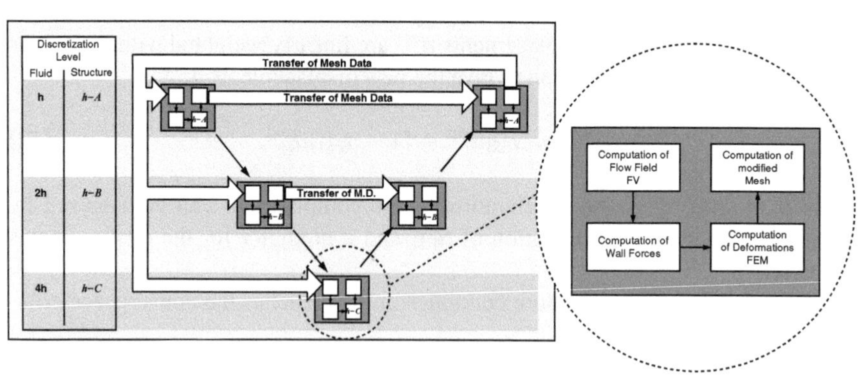

Fig. 2 Transfer of Mesh Movement Data for **Multi-Level Transfer**

From here, the interpolation to the finer grid levels starts. Now the meshes are the deformed meshes obtained during restriction. For the flow solver, the values are interpolated in the standard way of multi-grid up to the finest grid level. It must be pointed out, that there is no multi-grid algorithm applied for the structure part, and the grid levels on the structure side are not coupled.

3.3 Flow-Acoustics-Coupling

The coupling of flow and acoustics is done by inserting the flow induced acoustic sources into the acoustic equations 10. These sources can directly be calculated by the time derivation of the pressure in the flow field. The acoustics solver calculates the sound propagation. For the considered cases, the kinetic energy of the sound is too small that it can influence the incompressible flow field. Hence, the overall coupling for one time step contents four procedures:

1. Calculate the flow field, with its pressure field
2. Derivate the acoustic sources
3. Transfer the acoustic sources to the acoustics solver
4. Calculate the sound propagation.

While fluid and structure are connected via the surface of the structure, the flow-acoustics-coupling is a volume coupling. The computational effort of a volume coupling – including the data transfer, and the search of adjacent cells and interpolation – is much higher than for a surface coupling. Therefore, an integrated approach, in which the acoustics solver is integrated in the flow solver, is beneficial for saving computer resources.

In the presented approach Procedure 1 is done by the flow solver FASTEST with an LES, for Procedure 2 the calculated pressure field is differentiated with respect to t. The discretization for the acoustics solver and the flow solver is the same, thus the transfer in Procedure 3 is straightforward. For Procedure 4 the sound propagation is calculated on the basis of the linearized Euler equations (LEE) (Eq. (8)–(10)). A detailed description of the implementation can be found in [8, 9].

Acoustics and incompressible flow acts on different time and length scales. For low Mach number flows the speed of sound c is much larger than the speed of the flow, and the relevant length scales of turbulent flow are much smaller than the wave lengths of sound. Using the time step size, that results from the maximum time step size suitable for the explicit FV acoustics solver, the computation of the incompressible flow field would be very inefficient, since the time step would be unnecessarily small. For this reason we apply an uneven time coupling or "frozen fluid approach" (see Fig. 3a), so that only at each Nth acoustic time step the aeroacoustic source terms are updated and N acoustic time steps of the size Δt_{CAA} are performed within a single CFD time step of the size Δt_{CFD}, where

$$N = \frac{\Delta t_{CFD}}{\Delta t_{CAA}}. \tag{14}$$

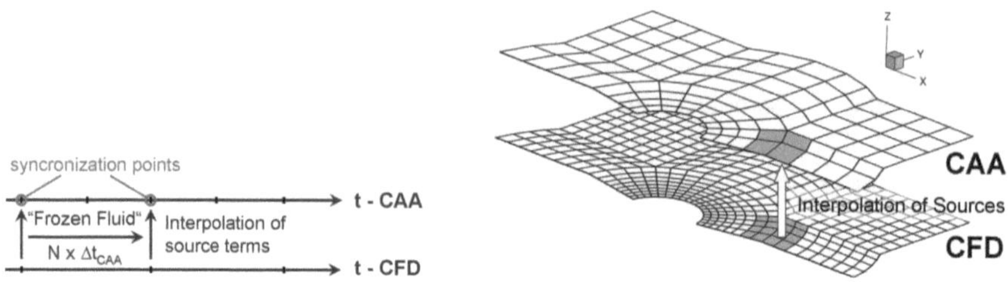

(a) Time stepping: data tranfer for CFD-CAA coupling.

(b) Interpolation from CFD-grid to acoustic grid using the multi-grid infra structure.

Fig. 3 Schematic view of different mesh sizes and data transfer, in time and space, for acoustics and flow solver

Fig. 4 Configuration of lid driven cavity test case

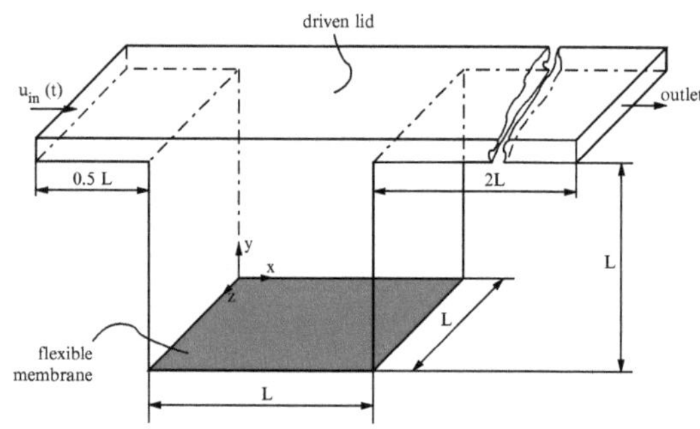

To consider the different requirements of the spatial discretization, the data infra structure of the geometric multi-grid method of FASTEST can be used: The flow field (velocity and pressure, Eq. (1)–(2)) is calculated on a fine grid level. On that grid the time derivatives of the pressure are computed and transferred as acoustic sources, by using the interpolation subroutines of the multi-grid infra structure, to a coarser grid level on which the acoustic propagation is computed. Figure 3b illustrates the grid dependencies.

4 Results

4.1 Fluid-Structure Interaction

As a test case, we consider a three dimensional lid driven cavity, as shown in Fig. 4 with $L = 1\text{m}$. At the bottom a flexible plate with thickness $t_{\text{mem}} = 0.1\text{m}$ is situated. The membrane is fixed at the edges and the lid moves with a time dependent lid velocity v_{lid} defined by

$$v_{\text{lid}}(t) = 0.5(1.5 - \cos(2\pi t / T_0)) \tag{15}$$

Table 1 Computing times in [s] and acceleration of single- and multi-grid methods

Grid set	Single-grid	Multi Grid	MG acceleration
coarse	2278	1202	1.9
medium	15297	1943	7.9
fine	260483	21445	12.2

with period $T_0 = 5$s. The inflow velocity corresponds to the lid velocity. The material parameters for the structure and the fluid are defined as: Young's modulus $E_s = 50{,}000\ \mathrm{N/m^2}$, Poisson ratio $\nu_s = 0.3$, structure density $\rho_s = 100\ \mathrm{kg/m^3}$, fluid dynamic viscosity $\mu_f = 0.01\ \mathrm{kg/(ms)}$, fluid density $\rho_f = 100\ \mathrm{kg/m^3}$ and static pressure $p_{stat} = 0.1$Pa. E_s and ν_s are related to the Lamé constants.

For the investigations, three successively refined fluid grids are considered: a *fine grid* with 159,744 control volumes, a *medium grid* with 19,968 control volumes, and a *coarse grid* with 2,496 control volumes. The multi-grid computations are done with three grid levels for each grid set. This means for the algorithm used, a factor of eight in the number of control volumes from one grid level to the next. The grid of the structure consists of 8-node brick elements (# of elements: *h-C*: $5 \times 1 \times 5$, *h-B*: $10 \times 2 \times 10$, *h-C*: $20 \times 4 \times 20$). The underrelaxation parameter is set to $\alpha_{FSI} = 0.8$ for all cases except the one involving the Aitken method. This value was ascertained to be a good choice for the used materials (see [6]).

For the Newmark time integration scheme the parameters are set to $\beta = 0.3$ and $\gamma = 0.6$ (see [19]).

First we investigate the multi-grid acceleration for the **Single-Level Transfer**, for this case the *h-C* discretization for the structure is used. Table 1 summarizes the computation times and accelerations for the single- and multi-grid method, spent for computing a time interval of $T = 1.5$ s with a time step size of $\Delta t = 0.1$ s. The computer employed was *Pentium 4 PC* with a clock rate of 2539 MHz.

One can obtain the typical acceleration by the multi-grid method increasing with the grid size.

Next, the different multi-grid variants are compared. The convergence behavior of **Single-Level Transfer** and **Multi-Level Transfer**, with fixed underrelaxation parameter $\alpha_{FSI} = 0.8$, is shown in Fig. 5.

The convergence behavior shows, that the stronger coupling with the *multi-level transfer* is a promising way for accelerating the FSI computations, when involving an effective coupling tool.

4.2 Flow and Acoustics

To investigate the parallel efficiency of the acoustics part of the solver and the efficiency of the discretization approach, computations for two different test cases are carried out. The fluid properties for both test cases are: fluid density $\rho =$

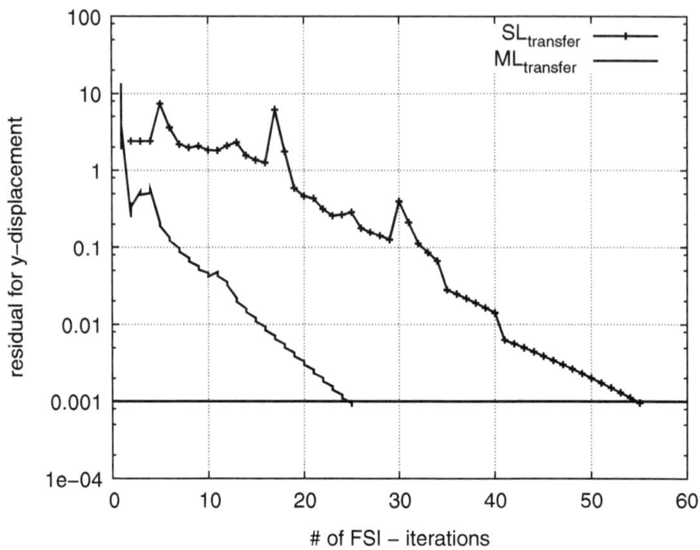

Fig. 5 FSI-convergence behavior of **Single-Level (SL)Transfer** and **Multi-Level (ML) Transfer**

1.225 kg/m^3 , hydrostatic pressure $p = 10^5$ Pa, speed of sound $c = 347$ m/s, kinematic viscosity $\mu = 1.8 \times 10^{-5}$kg/(m s).

The parallel efficiency and scalability of the implemented method is evaluated by using a generic test case which consists of a unit cube (1m^3) with non reflecting boundary conditions at all boundaries. The problem domain is discretized with a single block that can be split for domain decomposition when performing parallel computations for the evaluation of the parallel efficiency of the method. The numerical grid is equidistant in all directions. The number of CV can be reduced by coarsening the finest discretization with more than 16×10^6 CV.

Two numerical setups of the pressure pulse with 256^3 CV and 128^3 CV with different processor numbers up to 128 CPUs were performed. The computational domain is split into P domains for P CPUs with the same number of CV to obtain a load balancing of 100%. Further it was taken care to minimize the interface sizes for more than 2 CPUs for optimal communication between the processors. 100 acoustic time steps were calculated with an acoustic Courant number of approximately 0.4. Computing times, parallel speedup, and the parallel efficiency are evaluated for the performed simulations. The speedup S_P is defined as the ratio of the computing times obtained for one processor t_{1CPU} and for P CPUs t_{PCPUs}:

$$S_P = \frac{t_{1CPU}}{t_{PCPUs}}. \tag{16}$$

Further we define an alternative speedup S_P^* for more than 2 CPUs with the reference computing time t_{2CPUs} obtained for 2 CPUs:

$$S_P^* = \frac{2 \cdot t_{2CPUs}}{t_{PCPUs}}; \quad P \geq 2. \tag{17}$$

This allows to partly account for the communication latencies that may play a role for parallel computations if they are communication intense. The corresponding parallel efficiency E_P is defined as:

$$E_P = \frac{t_{1CPU}}{P \cdot t_{PCPUs}} \tag{18}$$

and

$$E_P^* = \frac{2 \cdot t_{2CPUs}}{P \cdot t_{PCPUs}}; \quad P \geq 2. \tag{19}$$

The numerical investigations are carried out on an IBM p575 system with 32 IBM Power6 Cores (4.7 GHz) per node. Figure 6a shows the computing times for both domain sizes together with the linear speedup for the reference computing times obtained for 1 and 2 CPUs. The very low speedup from 1 to 2 CPUs (compare also Fig. 6c and d) is remarkable and can be explained by communication latencies as well as the largest interface size. Further, it can be stated that latencies play a big role for the overall parallel efficiency of the presented method since the simple explicit scheme performs relatively few operations until a data exchange between two CPUs takes place.

Applying the alternative definition of the parallel speedup and the parallel efficiency (Eq. (17) and (19)) it can be seen, that for a wide range of CPU numbers a good parallel efficiency can be obtained for both domain sizes. For the smaller domain size with approximately 2×10^6 CV a superlinear speedup (respectively more than 100 % parallel efficiency) is obtained from 4 to 32 CPUs. This behavior can be explained by so called cache effects, where the CV number per processor becomes small enough so that a significant part of the problem can be stored in the internal CPU cache which is much faster than the RAM.

Another interesting fact is the change of the slope of the curves for 64 CPUs, where more than one node is incorporated in the computation. A further decline of efficiency can be seen for 128 CPUs where 4 nodes are used. For the smaller domain the overall computation time even increases from 64 to 128 CPUs.

In summary, it can be stated that the numerical scheme and its implementation show a good parallel scalability, even for large domains and a large number of processors.

With the second test case – a plate in the turbulent wake of a circular cylinder – the efficiency of the hierarchical discretization scheme is investigated. To give an impression of the turbulent structures around the test case, a component of the vorticity vector is shown in Fig. 7a. The interaction of the vortex shedding at the

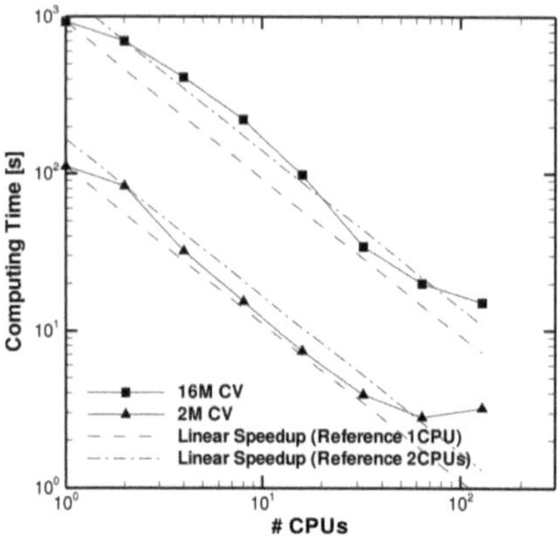

(a) Computing time

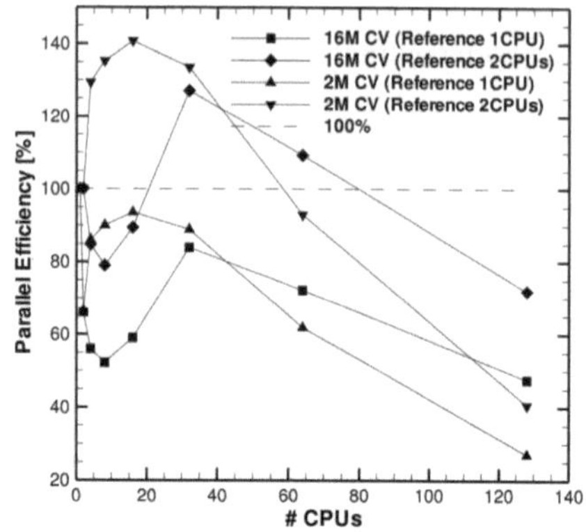

(b) Parallel efficiency

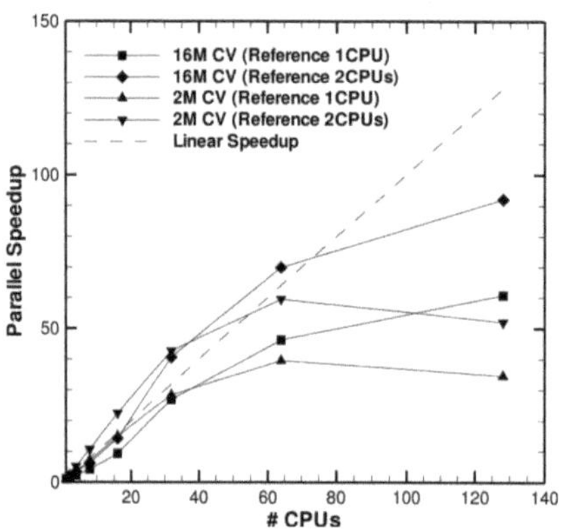

(c) Parallel speedup

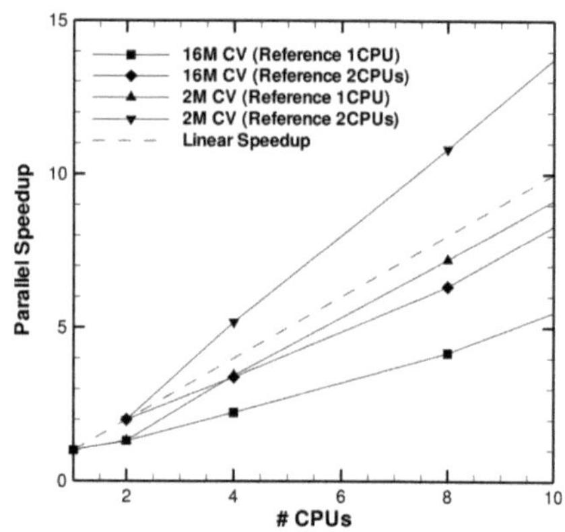

(d) Parallel speedup-detailed view

Fig. 6 Parallel efficiency of the acoustics solver

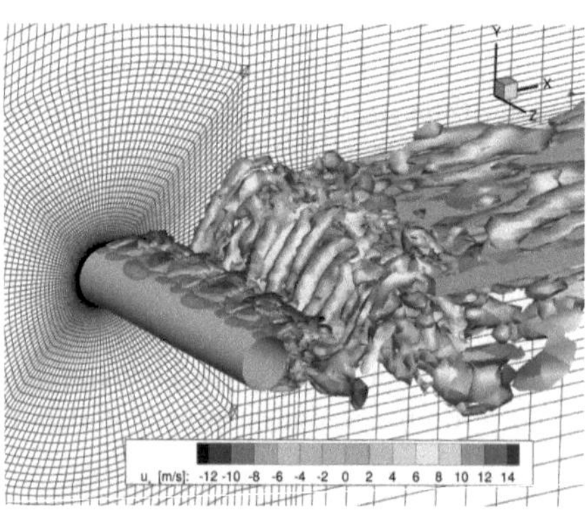

(a) Vortical structures in the vicinity. Component of the vorticity vector colored by the velocity component u_x in x-direction and numerical grid (each second grid line is shown).

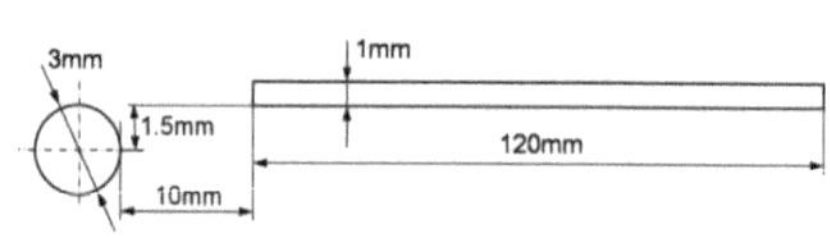

(b) Dimensions of the embedded structures

Fig. 7 Configuration, grid and simulation results of the cylinder-plate-configuration

Table 2 Numerical parameters for the aeroacoustic simulations

	CAA grid = CFD grid	1×coarsened CFD grid
# CV of CAA grid	6321408	790176
Δt_{CFD} [s]	1.0×10^{-6}	1.0×10^{-6}
$\#\Delta t_{CAA}$ within 1 CFD step	160	80
Δt_{CAA} [s]	1.0×10^{-8}	2.0×10^{-8}

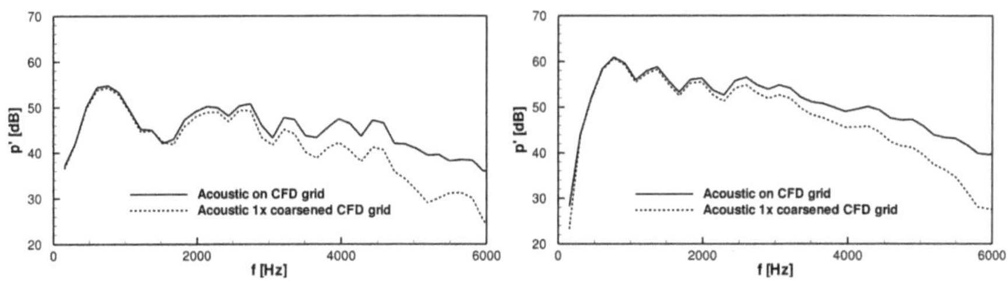

Fig. 8 Computed sound pressure spectra for two different monitoring positions and different acoustic grids

cylinder with the front edge of the plate leads to high time dependent pressure fluctuations and therefore high aeroacoustic sources. Experimental investigations and numerical studies (see [10]) show a significant increase of the sound pressure level compared to a single cylinder. This makes the test case very interesting as validation case for low Mach number aeroacoustics – as documented in [8].

Dimensions of the configuration are shown in Fig. 7b. The computational domain is of the size 1088 mm × 940 mm × 15 mm and the CFD grid consists of 6.32×10^6 CV. The simulations are carried out with 16 CPUs and a load balancing efficiency of 95.84%. The boundary layer around the cylinder is fully resolved whereas the boundary layer around the plate is not resolved. All walls are no slip walls for the flow and sonically hard walls for the acoustics. In spanwise direction periodic boundary conditions are applied. All other boundaries are non reflecting for acoustics and inlet respectively outlet boundaries for the flow.

LES are performed using the Smagorinsky model with the dynamic approach of Germano [4].

Two aeroacoustic simulations with two different grids for the acoustics were carried out: a simulation where CFD and CAA grid are identical and another simulation where the CAA grid is the 1×coarsened CFD grid. The numerical parameters for both simulations are summarized in Table 2.

Figure 8 shows the computed sound pressure spectra in two monitoring positions (80 mm above/below the center of the cylinder). Both spectra for the acoustics on the CFD grid and acoustics on the 1 × coarsened CFD grid are in good agreement for lower frequencies whereas for higher frequencies, that correspond with shorter wave lengths and therefore a lower resolution, the differences become larger. Nevertheless for many applications the differences in the spectra are acceptable, especially if the

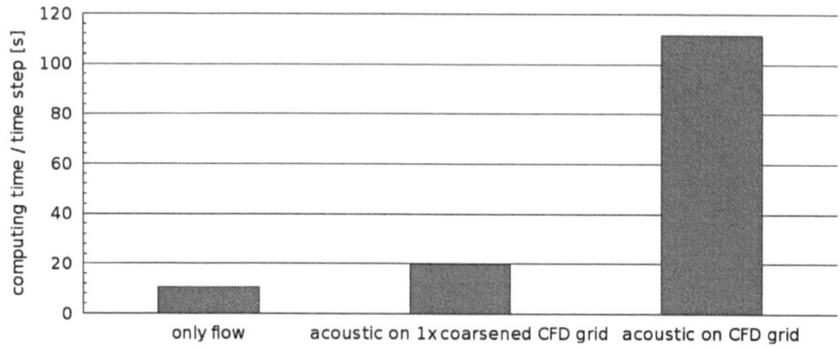

Fig. 9 Computing times for one time step and different acoustic grids in comparison to a pure CFD computation

focus lies in the qualitative behavior of different configurations during the design process. More results can be found in [8].

The computational times for both numerical setups are summarized in Fig. 9 in comparison to a pure CFD calculation. The aeroacoustic simulation where the acoustics are computed on the CFD mesh takes approximately 5 × longer than the simulation with the acoustics on the 1×coarsened CFD grid. The coupled aeroacoustic simulation with the coarser acoustic grid takes approximately 2 × longer than a pure CFD calculation of the same test case. It can be stated that an enormous benefit in terms of computing time can be achieved by applying the described hierarchical grid method. It should be mentioned that the differences between pure CFD and coupled simulations, as well as between the coupled simulations with different acoustic grids, strongly depend on the convergence behavior of the CFD simulation. For the presented simulation only 4 SIMPLE iterations were necessary to fulfill the convergence criterion. The differences between the simulations would be smaller for a worse convergence behavior since the computing time required for the acoustics stays constant for a given grid with the implemented explicit scheme.

5 Summary and Outlook

We presented a framework in which the full coupling of flow, structure, and acoustics is possible. Special emphasis has been given to the use of geometric multi-grid techniques for FSI and the use of the multi-grid infra structure in the flow solver for the reduction of computational effort.

Different possibilities for accelerating the FSI-simulations by the usage of multi-grid methods have been discussed. The results show a high benefit and motivate further investigations.

A numerical scheme for the simulation of aerodynamic noise caused by flows at low Mach numbers within and/or around complex geometries has been presented: the in-house finite volume flow solver FASTEST extended by a fully parallelized high resolution finite volume scheme for the linearized Euler equations. To speed up the coupled simulations a frozen fluid approach as well as a hierarchically grid based method for the aeroacoustic simulation has been proposed. The implemented scheme seems to be a very promising approach, especially for situations where the CFD and CAA domains are approximately of the same size.

Together, these approaches promise to constitute the basis for an efficient tool for simulation coupled fluid, solid, and acoustics phenomena.

Acknowledgement This research hasn't been possible without the support of DFG (DFG Forschergruppe 493 and Sonderforschungsbereich SFB 568).

References

1. Briggs, W.L., Henson, V.E., McCormick, S.F.: A Multigrid Tutorial. SIAM (2000)
2. Demirdzic, I., Peric, M.: Space conservation law in finite volume calculations of fluid flow. Int. J. Numer. Meth. Fluid. **8**(9), 1037–1050 (1988)
3. Ertem-Müller, S.: Numerical efficiency of implicit and explicit methods with multigrid for Large Eddy Simulation in complex geometries. PhD thesis, TU Darmstadt (2003)
4. Germano, M., Piomelli, U., Moin, P., Cabot, W.H.: A dynamic subgrid-scale eddy viscosity model. Phys. Fluids A **3**(7), 1760–1765, July (1991) LES 1.18
5. Heck, M.: Mehrgitterverfahren zur effizienten numerischen Simulation von Fluid-Struktur-Wechselwirkungen. PhD thesis, TU Darmstadt (2008)
6. Heck, M., Sternel, D.C., Schäfer, M., Yigit, S.: Influence of numerical and physical parameters on an implicit partitioned fluid-structure solver. In Proceedings of European Conference on CFD ECCOMAS (2006)
7. Institute of Numerical Methods in Mechanical Engineering, TU Darmstadt. *FASTEST-Manual*, 1 edition (2005)
8. Kornhaas, M.: Effiziente numerische Methoden zur Simulation aeroakustischer Probleme mit kleinen Machzahlen. PhD thesis, TU Darmstadt (2011)
9. Kornhaas, M., Sternel, D.C., Schäfer, M.: Efficiency investigation of a parallel hierarchical grid based aeroacoustic code for low mach numbers and complex geometries. In: Proceedings of V ECCOMAS CFD (2010)
10. Kornhaas, M., Winkler, M., Sternel, D.C., Becker, K., Schäfer, M., Kameier, F.: Les of cylinder plate configurations and corresponding aeroacoustic effects. In: Proceedings of 19ième Congrès Français de Mècanique (2009)
11. Leer, B.V.: Towards the ultimate conservative difference scheme iii. upstream-centered finite-difference schemes for ideal compressible flow. J. Comput. Phys. **23**, 263–275 (1977)
12. Lehnhäuser, T., Schäfer, M.: Improved linear interpolation practice for finite-volume schemes on complex grids. Int. J. Numer. Meth. Fluid. **38**, 625–645 (2002)
13. MpCCI. Mesh based parallel code coupling interface, user's guide V2.0. Fraunhofer Institut für Algorithmen und wissenschaftliches Rechnen, SCAI, Bonn (2004)
14. Schäfer, M., Sternel, D.C., Becker, G., Pironkov, P.: Efficient numerical simulation and optimization of fluid-structure interaction. In: Fluid Structure Interaction II: Modelling, Simulation, Optimization, vol. 73, pp. 131–158. Springer (2010)
15. Shen, W.Z., Sørensen, J.N.: Aeroacoustic modelling of low-speed flows. Theor. Comput. Fluid Dynam. **13**, 271–289 (1999)
16. Shen, W.Z., Sørensen, J.N.: Comment on the aeroacoustic formulation of hardin and pope. AIAA J. **37**(1), 141–143 (1999)
17. Sternel, D.C., Schäfer, M., Heck, M., Yigit, S.: Efficiency and accuracy of fluid-structure interaction simulations using an implicit partitioned approach. Comput. Mech. **43**(1), 103–113 12 (2008)

18. Taylor, R.L.: FEAP - A Finite Element Analysis Program - Version 7.5 User Manual. University of California (2003)
19. Wood, W.L.: Practical Time-stepping Schemes. Clarendon Press, Oxford (1990)
20. Yigit, S., Schäfer, M., Heck, M.: Grid movement techniques and their influence on laminar fluid structure interaction computations. J. Fluid Struct. **24**(6), 819–832 (2008)
21. Yigit, S., Sternel, D., Schäfer, M.: Efficiency of fluid-structure interaction simulations with adaptive underrelaxation and multigrid acceleration. Int. J. Multiphysics **1**, 85–99 (2007)

LIKWID: Lightweight Performance Tools

Jan Treibig, Georg Hager, and Gerhard Wellein

Abstract Exploiting the performance of today's microprocessors requires intimate knowledge of the microarchitecture as well as an awareness of the ever-growing complexity in thread and cache topology. LIKWID is a set of command line utilities that addresses four key problems: Probing the thread and cache topology of a shared-memory node, enforcing thread-core affinity on a program, measuring performance counter metrics, and microbenchmarking for reliable upper performance bounds. Moreover, it includes an mpirun wrapper allowing for portable thread-core affinity in MPI and hybrid MPI/threaded applications. To demonstrate the capabilities of the tool set we show the influence of thread affinity on performance using the well-known OpenMP STREAM triad benchmark, use hardware counter tools to study the performance of a stencil code, and finally show how to detect bandwidth problems on ccNUMA-based compute nodes.

1 Introduction

Today's multicore x86 processors bear multiple complexities when aiming for high performance. Conventional performance tuning tools like Intel VTune, OProfile, CodeAnalyst, OpenSpeedshop, etc., require a lot of experience in order to get sensible results. For this reason they are usually unsuitable for the scientific users, who would often be satisfied with a rough overview of the performance properties of their application code. Moreover, advanced tools often require kernel patches and additional software components, which make them unwieldy and bug-prone. Additional confusion arises with the complex multicore, multicache, multisocket

J. Treibig (✉) · G. Hager · G. Wellein
Erlangen Regional Computing Center (RRZE), Friedrich-Alexander Universität Erlangen-Nürnberg, Martensstr. 1, D-91058 Erlangen, Germany
e-mail: jan.treibig@rrze.uni-erlangen.de; georg.hager@rrze.uni-erlangen.de; gerhard.wellein@rrze.uni-erlangen.de

C. Bischof (eds.), *Competence in High Performance Computing 2010*,
DOI 10.1007/978-3-642-24025-6_14,

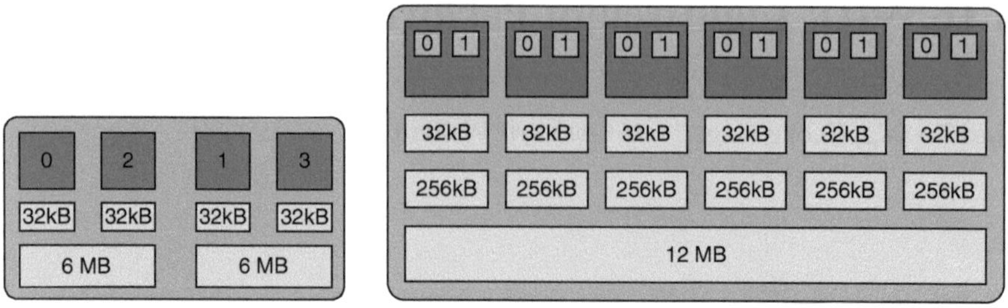

(a) Core 2 Quad (b) Nehalem EP Westmere

Fig. 1 Cache and thread topology of Intel Core 2 Quad and Nehalem EP Westmere processors

structure of modern systems (see Fig. 1); users are all too often at a loss about how hardware thread IDs are assigned to resources like cores, caches, sockets and NUMA domains. Moreover, the technical details of how threads and processes are bound to those resources vary strongly across compilers and MPI libraries.

LIKWID ("Like I Knew What I'm Doing") is a set of easy to use command line tools to support optimization. It is targeted towards performance-oriented programming in a Linux environment, does not require any kernel patching, and is suitable for Intel and AMD processor architectures. Multithreaded and even hybrid shared/distributed-memory parallel code is supported. LIKWID comprises the following tools:

- `likwid-features` can display and alter the state of the on-chip hardware prefetching units in Intel x86 processors.
- `likwid-topology` probes the hardware thread and cache topology in multi-core, multisocket nodes. Knowledge like this is required to optimize resource usage like, e.g., shared caches and data paths, physical cores, and ccNUMA locality domains in parallel code.
- `likwid-perfCtr` measures performance counter metrics over the complete runtime of an application or, with support from a simple API, between arbitrary points in the code. Although it is possible to specify the full, hardware-dependent event names, some predefined event sets simplify matters when standard information like memory bandwidth or Flop counts is needed.
- `likwid-pin` enforces thread-core affinity in a multi-threaded application "from the outside," i.e., without changing the source code. It works with all threading models that are based on POSIX threads, and is also compatible with hybrid "MPI+threads" programming. Sensible use of likwid-pin requires correct information about thread numbering and cache topology, which can be delivered by likwid-topology (see above).
- `likwid-mpirun` allows to pin a pure MPI or hybrid MPI/threaded application to dedicated compute resources in an intuitive and portable way.
- `likwid-bench` is a microbenchmarking framework allowing rapid prototyping of small assembly kernels. It supports threading, thread and memory

placement, and performance measurement. `likwid-bench` comes with a wide range of typical benchmark cases and can be used as a stand-alone benchmarking application.

Although the six tools may appear to be partly unrelated, they solve the typical problems application programmers encounter when porting and running their code on complex multicore/multisocket environments. Hence, we consider it a natural idea to provide them as a single tool set.

This paper is organized as follows. Section 2 describes two of the tools in some detail and gives hints for typical use. Section 3 demonstrates the use of LIKWID in three different case studies, and Sect. 4 gives a summary and an outlook to future work.

2 Tools

LIKWID only supports x86-based processors. Given the strong prevalence of those architectures in the HPC market (e.g., 90% of all systems in the latest Top 500 list are of x86 type) we do not consider this a severe limitation. In other areas like, e.g., workstations or desktops, the x86 dominance is even larger.

An important concept shared by all tools in the set is logical numbering of compute resources inside so-called *thread domains*. Under the Linux OS, hardware threads in a compute node are numbered according to some scheme that heavily depends on the BIOS and kernel version, and which may be unrelated to natural topological units like cache groups, sockets, etc. Since users naturally think in terms of topological structures, LIKWID introduces a simple and yet flexible syntax for specifying processor resources. This syntax consists of a prefix character and a list of logical IDs, which can also include ranges. The following domains are supported:

Node	`N`
Socket	`S[0-9]`
Last level shared cache	`C[0-9]`
NUMA domain	`M[0-9]`

Multiple ID lists can be combined, allowing a flexible numbering of compute resources. To indicate, e.g., the first two cores of NUMA domains 1 and 3, the following string can be used: `M0:0,1@M2:0,1`.

In the following we describe two of the six tools in more detail. A thorough documentation of all tools apart from the man pages is found on the WIKI pages on the LIKWID homepage [1].

2.1 *likwid-perfctr*

Hardware-specific optimization requires an intimate knowledge of the microarchitecture of a processor and the characteristics of the code. While many problems can be solved with profiling, common sense, and runtime measurements, additional information is often useful to get a complete picture.

Performance counters are facilities to count hardware events during code execution on a processor. Since this mechanism is implemented directly in hardware there is no overhead involved. All modern processors provide hardware performance counters. They are attractive for application programmers, because they allow an in-depth view on what happens on the processor while running applications. As shown below, `likwid-perfctr` has practically zero overhead since it reads performance metrics at predefined points. It does not support statistical counter sampling. At the time of writing, `likwid-perfctr` runs on all current x86-based architectures.

The probably best known and widespread existing tool is the PAPI library [5, 6]. A lot of research is targeted towards using hardware counter data for automatic analysis and detecting potential performance bottlenecks [2–4]. However, those solutions are often too unwieldy for the common user, who would prefer a quick overview as a first step in performance analysis. A key design goal for `likwid-perfctr` was ease of installation and use, minimal system requirements (no additional kernel modules and patches), and – at least for basic functionality – no changes to the user code. A prototype for the development of `likwid-perfctr` is the SGI tool "perfex," which was available on MIPS-based IRIX machines as part of the "SpeedShop" performance suite. Cray provides a similar, PAPI-based tool (craypat) on their systems [8]. `likwid-perfctr` is a dedicated command line tool for programmers, allowing quick and flexible measurement of hardware performance counters on x86 processors, and is available as open source. It allows simultaneous measurements on multiple cores. Events that are shared among the cores of a socket (this pertains to the "uncore" events on Core i7-type processors) are supported via "socket locks," which enforce that all uncore event counts are assigned to one thread per socket. Events are specified on the command line, and the number of events to count concurrently is limited by the number of performance counters on the CPU. These features are available without any changes in the user's source code. A small instrumentation ("marker") API allows one to restrict measurements to certain parts of the code (named regions) with automatic accumulation over all regions of the same name. An important difference to most existing performance tools is that event counts are strictly core-based instead of process-based: Everything that runs and generates events on a core is taken into account; no attempt is made to filter events according to the process that caused them. The user is responsible for enforcing appropriate affinity to get sensible results. This could be achieved with `likwid-perfctr` itself or alternatively via `likwid-pin` (see below for more information):

```
$ likwid-perfctr -C S0:0 \
   -g SIMD_COMP_INST_RETIRED_PACKED_DOUBLE:PMC0,\
      SIMD_COMP_INST_RETIRED_SCALAR_DOUBLE:PMC1 ./a.out
```

(See below for typical output in a more elaborate setting.) In this example, the computational double precision packed and scalar SSE retired instruction counts on an Intel Core 2 processor are assigned to performance counters 0 and 1 and measured on the first core (ID 0) of the first socket (domain S0) over the duration of `a.out`'s runtime. As a side effect, it becomes possible to use `likwid-perfctr` as a monitoring tool for a complete shared-memory node, just by specifying all cores for measurement and, e.g., "`sleep`" as the application:

```
$ likwid-perfctr -c N:0-7 \
   -g SIMD_COMP_INST_RETIRED_PACKED_DOUBLE:PMC0,\
      SIMD_COMP_INST_RETIRED_SCALAR_DOUBLE:PMC1 \
         sleep 1
```

Apart from naming events as they are documented in the vendor's manuals, it is also possible to use preconfigured *event sets* (groups) with derived metrics. This provides a simple abstraction layer in cases where standard information like memory bandwidth, Flops per second, etc., is sufficient:

```
$ likwid-perfctr -C N:0-3 -g FLOPS_DP  ./a.out
```

The event groups are partly inspired from a technical report published by AMD [7], and all supported groups can be obtained by using the `-a` command line switch. We try to provide the same preconfigured event groups on all supported architectures, as long as the native events support them. This allows the beginner to concentrate on the useful information right away, without the need to look up events in the manuals (similar to PAPI's high-level events). In the usage scenarios described so far there is no interference of `likwid-perfctr` while user code is being executed, i.e., the overhead is very small (apart from the unavoidable API call overhead in marker mode).

The following example illustrates the use of the marker API in a serial program with two named regions ("`Main`" and "`Accum`"):

```
#include <likwid.h>
...
int coreID = likwid_processGetProcessorId();
likwid_markerInit(numberOfThreads,numberOfRegions);
int MainId  = likwid_markerRegisterRegion("Main");
int AccumId = likwid_markerRegisterRegion("Accum");

likwid_markerStartRegion(0, coreID);
// measured code region "Main" here
likwid_markerStopRegion(0, coreID, MainId);

for (j = 0; j < N; j++) {
   likwid_markerStartRegion(0, coreID);
   // measured code region "Accum" here
   likwid_markerStopRegion(0, coreID, AccumId);
}
likwid_markerClose();
```

Event counts are automatically accumulated on multiple calls. Nesting or partial overlap of code regions is not allowed. The API requires specification of a thread

ID (0 for one process only in the example) and the core ID of the thread/process. The LIKWID API provides simple functions to determine the core ID of processes or threads. The following listing shows the shortened output of `likwid-perfctr` after measurement of the `FLOPS_DP` event group on four cores of an Intel Core 2 Quad processor in marker mode with two named regions ("`Init`" and "`Benchmark`," respectively):

```
$ likwid-perfCtr -c 0-3 -g FLOPS_DP -m ./a.out
-------------------------------------------------------------
CPU type:       Intel Core 2 45nm processor
CPU clock:      2.83 GHz
-------------------------------------------------------------
Measuring group FLOPS_DP
-------------------------------------------------------------
Region: Init
+-------------------------------------+--------+--------+--------+--------+
|                Event                | core 0 | core 1 | core 2 | core 3 |
+-------------------------------------+--------+--------+--------+--------+
|          INSTR_RETIRED_ANY          | 313742 | 376154 | 355430 | 341988 |
|        CPU_CLK_UNHALTED_CORE        | 217578 | 504187 | 477785 | 459276 |
. . .
+-------------------------------------+--------+--------+--------+--------+
+-------------+-------------+-------------+-------------+-------------+
|   Metric    |   core 0    |   core 1    |   core 2    |   core 3    |
+-------------+-------------+-------------+-------------+-------------+
| Runtime [s] | 7.67906e-05 | 0.000177945 | 0.000168626 | 0.000162094 |
|     CPI     |  0.693493   |   1.34037   |   1.34424   |   1.34296   |
| DP MFlops/s |  0.0130224  | 0.00561973  | 0.00593027  | 0.00616926  |
+-------------+-------------+-------------+-------------+-------------+
Region: Benchmark
+---------------------+-------------+-------------+-------------+-------------+
|        Event        |   core 0    |   core 1    |   core 2    |   core 3    |
+---------------------+-------------+-------------+-------------+-------------+
| INSTR_RETIRED_ANY   | 1.88024e+07 | 1.85461e+07 | 1.84947e+07 | 1.84766e+07 |
. . .
|     CPI     |  1.52023  |  1.52252   |  1.52708   |  1.52661   |
| DP MFlops/s |  1624.08  |  1644.03   |  1643.68   |   1645.8   |
+-------------+-----------+------------+------------+------------+
```

Note that the `INSTR_RETIRED_ANY` and `CPU_CLK_UNHALTED_CORE` events are always counted (using two nonassignable "fixed counters" on the Core 2 architecture), so that the derived `CPI` metric ("cycles per instruction") is easily obtained.

2.2 *likwid-pin*

Thread/process affinity is vital for performance. If topology information is available, it is possible to "pin" threads according to the application's resource requirements like bandwidth, cache sizes, etc. Correct pinning is even more important on processors supporting SMT, where multiple hardware threads share resources on a single core. `likwid-pin` supports thread affinity for all threading models that are based on POSIX threads, which includes most OpenMP implementations. By overloading the `pthread_create` API call with a shared library wrapper, each thread can be pinned in turn upon creation, working through a list of core IDs. This list, and possibly other parameters, are encoded in environment variables that are evaluated when the library wrapper is first called. `likwid-pin` simply starts the user application with the library preloaded.

This architecture is illustrated in Fig. 2. No code changes are required, but the application must be dynamically linked. This mechanism is independent of

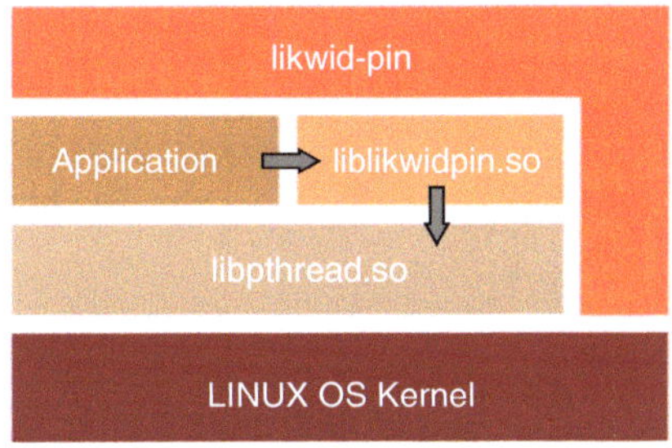

Fig. 2 Basic architecture of likwid-pin

the processor architecture, but the way the compiled code creates application threads must be taken into account: For instance, the Intel OpenMP implementation always runs `OMP_NUM_THREADS` threads but uses the first newly created thread as a management thread, which should not be pinned. This knowledge must be communicated to the wrapper library. The following example shows how to use `likwid-pin` with an OpenMP application compiled with the Intel compiler:

```
$ export OMP_NUM_THREADS=4
$ likwid-pin -c N:0-3 -t intel ./a.out
```

In general, `likwid-pin` can be used as a replacement for the `taskset` tool, which cannot pin threads individually. Currently, POSIX threads, Intel OpenMP, and GNU (gcc) OpenMP are supported directly, and the latter is assumed as the default if the `-t` option is not used. A bit mask can be specified to identify the management threads for cases not covered by the available parameters to the `-t` option. Moreover, `likwid-pin` can also be employed for hybrid programs that combine MPI with some threading model, if the MPI process startup mechanism establishes a Linux cpuset for every process.

The big advantage of `likwid-pin` is its portable approach to the pinning problem, since the same tool can be used for most applications, compilers, MPI implementations, and processor types. In Sect. 3.1 the usage model is analyzed in more detail on the example of the STREAM triad.

3 Case Studies

3.1 Case Study 1: Influence of Thread Topology on STREAM Triad Performance

To illustrate the general importance of thread affinity we use the well known OpenMP STREAM triad on an Intel Westmere dual-socket system. Intel Westmere is a hexacore design based on the Nehalem architecture and supports two SMT threads per physical core. The Intel C compiler version 11.1 was used with

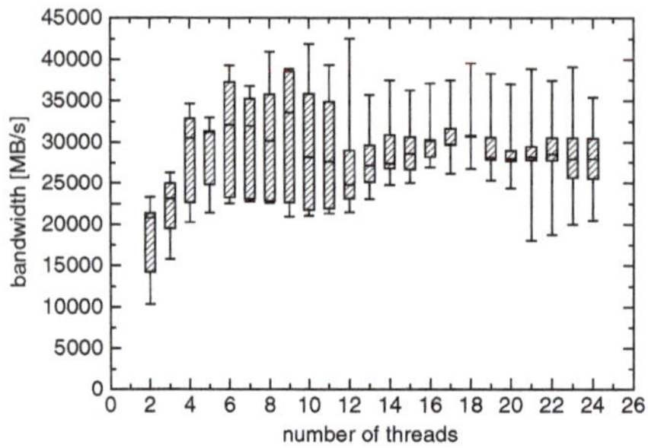

(a) not pinned

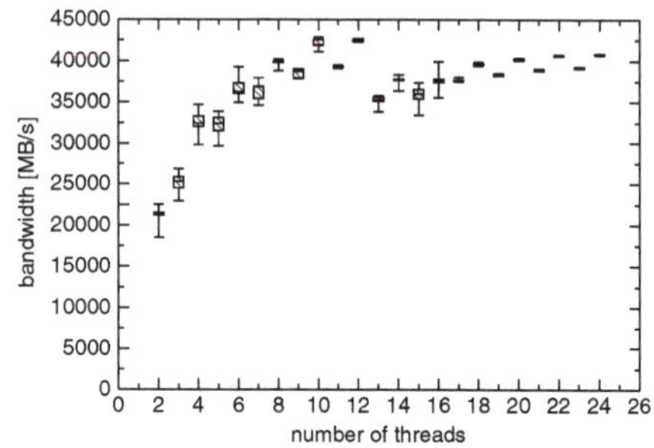

(b) pinned

Fig. 3 STREAM triad test run with the Intel C compiler on a dual-socket Intel Westmere system (six physical cores per socket). In (**a**) threads are not pinned and the Intel pinning mechanism is disabled. In (**b**) the application is pinned such that threads are equally distributed on the sockets to utilize the memory bandwidth in the most effective way. Moreover, the threads are first distributed over physical cores and then over SMT threads

options `-openmp -O3 -xSSE4.2 -fno-fnalias`. Intel compilers support thread affinity only if the application is executed on Intel processors. The functionality of this topology interface is controlled by setting the environment variable `KMP_AFFINITY`. In our tests `KMP_AFFINITY` was set to `disabled`. For the case of the STREAM triad on these ccNUMA architectures the best performance is achieved if threads are equally distributed across the two sockets.

Figure 3 shows the results. The non-pinned case shows a large variance in performance especially for the smaller thread counts where the probability is large that only one socket is used. With larger thread counts there is a high probability that both sockets are used, still there is also a chance that cores are oversubscribed, which reduces performance. The pinned case consistently shows high performance throughout. It is apparent that the SMT threads of Westmere increase the chance of different threads fighting for common resources.

3.2 Case Study 2: Monitoring the Performance of a Lattice Boltzmann Fluid Solver

To demonstrate the daemon mode option of `likwid-perfctr` an MPI-parallel Lattice Boltzmann fluid solver was analyzed on a Intel Nehalem quad-core system (Fig. 4). The daemon mode of `likwid-perfctr` allows time-resolved measurements of counter values and derived metrics in performance groups. It is used as follows:

```
$ likwid-perfctr -c S0:0-3 -g FLOPS_DP -d 800ms
```

This command measures the performance group `FLOPS_DP` on all physical cores of the first socket, with an interval of 800 ms between samples. `likwid-perfctr`

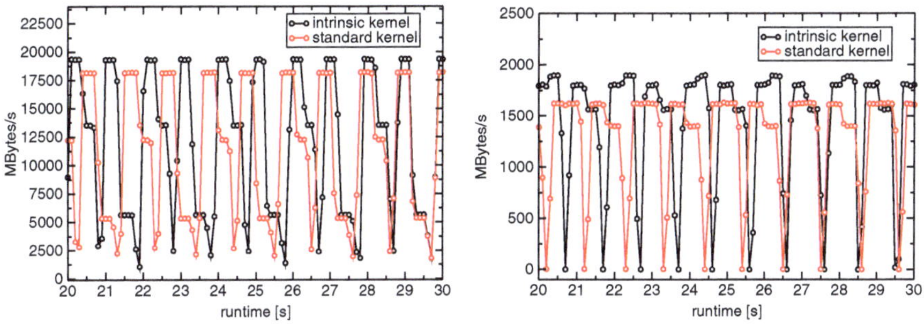

(a) Memory bandwidth (b) Floating point performance

Fig. 4 Time-resolved results for the iteration phase of an MPI-parallel Lattice Boltzmann solver on one socket (four cores) of an Intel Nehalem compute node. The compute performance in MFlops/s (**a**) and the memory bandwidth in MBytes/s (**b**) are shown over a duration of 10 sec, comparing two versions of the computational kernel (standard C versus SIMD intrinsics)

will only read out the hardware monitoring counters and print the difference between the current and the previous measurement. Therefore, the overhead is kept to a minimum. For this analysis the performance groups `FLOPS_DP` and `MEM` were used.

3.3 Case Study 3: Detecting ccNUMA Problems on Modern Compute Nodes

Many performance problems in shared memory codes are caused by an inefficient use of the ccNUMA memory organization on modern compute nodes. CcNUMA technology achieves scalable memory size and bandwidth at the price of higher programming complexity: The well-known locality and contention problems can have a large impact on the performance of multithreaded memory-bound programs if parallel first touch placement is not used on initialization loops, or is not possible for some reason [9].

`likwid-perfctr` supports the developer in detecting NUMA problems with two performance groups: `MEM` and `NUMA`. While on some architectures like, e.g., newer Intel systems, all events can be measured in one run using the `MEM` group, a separate group (`NUMA`) is necessary on AMD processors.

The example in Fig. 5 shows results for a memory copy benchmark, which is part of `likwid-bench`. Since `likwid-bench` allows easy control of thread and data placement it is well suited to demonstrate the capabilities of `likwid-perfctr` in detecting NUMA problems. Here, `likwid-perfctr` was used as follows:

```
$ likwid-perfctr -c S0:0@S1:0 -g MEM ./a.out
```

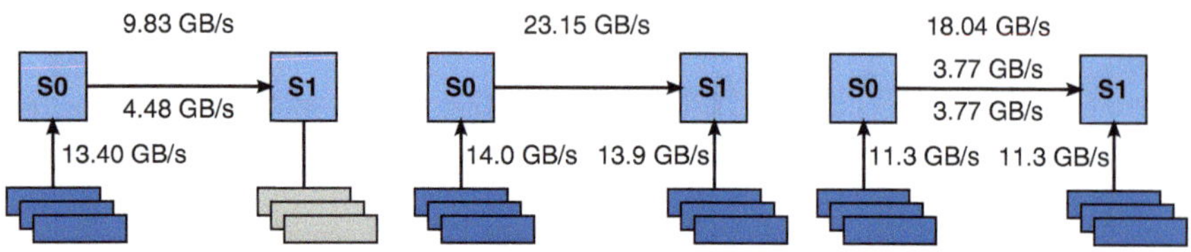

(a) Sequential initialization (b) First touch policy (c) Interleave policy

Fig. 5 NUMA problems reproduced with `likwid-bench` on the example of a memory copy benchmark on an Intel Nehalem dual-socket quad-core machine. The bandwidth on the top is the effective total application bandwidth as measured by `likwid-bench` itself. The other bandwidth values are from `likwid-perfctr` measurements in wrapper mode

The relevant output for the derived metrics could look like this:

```
+-----------------------------+-----------+----------+
|           Metric            |  core 0   |  core 4  |
+-----------------------------+-----------+----------+
|        Runtime [s]          |  4.71567  | 0.138517 |
|            CPI              |  16.4815  | 0.605114 |
| Memory bandwidth [MBytes/s] |  6.9273   | 6998.71  |
|  Remote Read BW [MBytes/s]  | 0.454106  | 4589.46  |
| Remote Write BW [MBytes/s]  | 0.0705132 | 2289.32  |
|    Remote BW [MBytes/s]     | 0.524619  | 6878.78  |
+-----------------------------+-----------+----------+
```

All threads were executed on socket zero, as can be seen from the runtime which is based on the `CPU_CLK_UNHALTED_CORE` metric. All program data originated from socket one since there is practically no local memory bandwidth. Hence, all bandwidth on socket one came from the remote socket. Figure 5a shows the results for sequential data initialization on one socket; the overall bandwidth is 9.83 GB/s. Figure 5b shows the case with correct first touch data placement on both sockets. The effective bandwidth is 23.15 GB/s, and the scalable ccNUMA system is used in the most efficient way. If an application cannot be easily changed to make use of the first touch memory policy, a viable compromise is often to switch to automatic round-robin page placement across a set of NUMA domains, or *interleaving*. `likwid-pin` can enforce interleaving for all NUMA domains included in a threaded run. This can be achieved with the `-i` option:

```
$ likwid-pin -c S0:0-3@S1:0-3 -t intel -i ./a.out
```

Figure 5c reveals that the memory bandwidth achieved with interleaving policy, while not as good as with correct first touch, is still much larger than the bandwidth of case (a) with all data in one NUMA domain.

4 Conclusion and Future Plans

LIKWID is a collection of command line applications supporting performance-oriented software developers in their effort to utilize today's multicore processors in an effective manner. LIKWID does not try to follow the trend to provide yet another complex and sophisticated tooling environment, which would be difficult to set up and would overwhelm the average user with large amounts of data. Instead it tries to make the important functionality accessible with as few obstacles as possible. The focus is put on simplicity and low overhead. `likwid-topology` and `likwid-pin` enable the user to account for the influence of thread and cache topology on performance and pin their application to physical resources in all possible scenarios with one single tool and no code changes. The usage of `likwid-perfctr` was demonstrated on two examples. LIKWID is open source and released under GPL2. It can be downloaded at `http://code.google.com/p/likwid/`.

Future plans include applying the philosophy of LIKWID to other areas like, e.g., profiling (also on the assembly level). Emphasis will also be put on a further improvement with regard to usability. It is also planned to port parts of LIKWID to the Windows operating system. An ongoing effort is to add support for present and upcoming architectures like, e.g., the Intel Sandy Bridge microarchitecture.

Acknowledgment We are indebted to Intel Germany for providing test systems and early access hardware for benchmarking. A special acknowledgment goes to Michael Meier, who had the basic idea for `likwid-pin`, implemented the prototype, and provided many useful thoughts in discussions. This work was supported by the Competence Network for Scientific and Technical High Performance Computing in Bavaria (KONWIHR) under the project "OMI4papps."

References

1. *Homepage of LIKWID tool suite* http://code.google.com/p/likwid/
2. Jost, G., Haoqiang, J., Labarta, J., Gimenez, J., Caubet, J.: Performance analysis of multilevel parallel applications on shared memory architectures. Proceedings of the Parallel and Distributed Processing Symposium (2003)
3. Gerndt, M., Kereku, E.: Automatic Memory Access Analysis with Periscope. ICCS '07: Proceedings of the 7th international conference on Computational Science pp. 847–854 (2007)
4. Gerndt, M., Fürlinger, K., Kereku, E.: Periscope: Advanced Techniques for Performance Analysis. PARCO, pp. 15–26 (2005)
5. Terpstra, D., Jagode, H., You, H., Dongarra, J.: Collecting Performance Data with PAPI-C. Proceedings of the 3rd Parallel Tools Workshop, Springer, Dresden, Germany (2010)
6. Browne, S., Deane, C., Ho, G., Mucci, P.: PAPI: A Portable Interface to Hardware Performance Counters. Proceedings of Department of Defense HPCMP Users Group Conference, June (1999)
7. Drongowski, P.J.: Basic Performance Measurements for AMD Athlon 64, AMD Opteron and AMD Phenom Processors. Technical Note, Advanced Micro Devices, Inc. Boston Design Center, September (2008)
8. DeRose, L., Homer, B., Johnson, D.: Detecting application load imbalance on high end massively parallel systems. Euro-Par 2007 Parallel Processing Conference, pp. 150–159 (2007)
9. Hager, G., Wellein, G.: Introduction to High Performance Computing for Scientists and Engineers. CRC Press, ISBN 978-1439811924, July (2010)

Scalability Issues of In-Situ Visualization in Parallel Simulation of Unsteady Flows

Michael Vetter and Stephan Olbrich

Abstract Grand challenge applications of 3-dimensional high resolution unsteady computational fluid dynamics result in huge amounts of data. To avoid significant bottlenecks of the storage and communication resources, efficient techniques for data extraction and preprocessing at the source have been realized in the parallel, network-distributed process chain called DSVR. Here the 3D data extraction is implemented as a parallel library and can be done in-situ during the numerical simulations, which avoids the storage of raw data for visualization. In this work we present, evaluate, and compare three techniques of parallel in-situ pathline extraction in distributed memory architectures. The gain in parallel scalability is achieved by an innovative trade-off between parallelization of partial tasks and asynchronous execution of suited serialized tasks. It has been shown that advanced parallelization schemes increase the scalability significantly.

1 Introduction

Computational Fluid Dynamics (CFD) is a major field of interest in recent scientific simulation applications. Coming along with increasing compute power of modern massively parallel compute clusters, these applications are highly parallelized, too, increasing resolution in space and time. This way the simulations of CFD models reach accuracy in the order of 10^{10} grid cells over 10^4 time steps, which results in a data volume of up to about 1 PByte.

For a better understanding and evaluation of the obtained data appropriate visualization becomes essential. A classical approach for the visualization of unsteady flows is the extraction of pathlines, representing the traces of virtual particles.

M. Vetter (✉) · S. Olbrich
Scientific Visualization and Parallel Processing, Regional Computing Center (RRZ), University of Hamburg
e-mail: michael.vetter@rrz.uni-hamburg.de; stephan.olbrich@rrz.uni-hamburg.de

C. Bischof (eds.), *Competence in High Performance Computing 2010*,
DOI 10.1007/978-3-642-24025-6_15,

In these graphics based approaches for flow field visualization one of the most discussed problems is the initial placement of these lines. This is normally solved by calculating adequate seed points via different techniques of numerical flow field analysis or distributing pathlines homogeneously on a planar surface or in 3-dimensional space. But for unsteady high-resolution flow fields with huge amounts of raw data up to 1 PByte the flow analysis cannot be done interactively. Distributing pathlines homogeneously is not an alternative because it leads to visual cluttering.

For reasons of efficiency this visualization task is moved to high performance computers more and more. But due to Amdahl's Law, the parallel speed-up is limited to the reciprocal value of the sequential fraction of the computation. For this reason, efficient utilization for the visualization can only be achieved if also the data extraction and visual data analysis components of the complete process chain are parallelized adequately.

This leads to the need of highly scalable parallel algorithms for the flow field visualization. Parallelization of pathline extraction can be done easily by decomposing the pathlines equally or as in-element method, where the raw data and the pathlines are allocated by domain decomposition. Though for the first approach the raw data has to be available on every processing node, due to the bottlenecks within storage and bandwidth, this is limited to shared memory architecture.

In-situ visualization provides a method to avoid the storage bottleneck. Here the raw data which the simulation produces is not stored but mapped directly to geometric 3D objects. The regard of the entire process chain was introduced in [5].

To take these challenges a distributed visualization environment has been implemented in the *DSVR Framework* avoiding the storage of raw data at all due to its in-situ visualization capability. In this article we present and compare three different approaches for efficient in-situ pathline extraction based on the in-element method within this framework. After giving an overview on the related work in Sect. 2, in Sect. 3 the DSVR framework is introduced in detail before in Sect. 4 the new developed algorithms are described. In Sect. 5 the algorithms are evaluated.

2 Related Work

Techniques for flow field visualization typically used for CFD data can be classified in *direct flow visualization* like drawing arrows for each vector stored in the field, *texture-based flow visualization* like LIC, *geometric flow visualization*, which consists of particle tracing using numerical integration, and *feature-based flow visualization*, where the flow field is analyzed for so-called "critical points". Overviews of these classifications and typical techniques are given by [7, 10, 13]. In this article we will concentrate on pathline extraction.

The huge amount of raw data calculated by recent CFD applications cannot be loaded into the memory of a single PC or cluster node. Also the parallelization of the visualization tasks does not help out. So out-of-core approaches for particle tracing had been introduced and enhanced by [1] and [3] for example. These out-of-core approaches avoid the memory bottleneck by loading just parts of the raw data into the memory of the compute nodes.

Within post-processing scenarios in-element approaches like the Parallel Particle Tracking Framework [2] for shared memory systems or the hierarchical distribution of raw data for distributed memory systems as introduced in [16, 17] become a way out of the bandwidth and memory bottleneck, but do not solve the storage problem.

For post-processing streamlines from large datasets recently were discussed three parallelization strategies [14]: by domain decomposition, by line distribution, and a new hybrid approach combining these two.

VIRACOCHA [4] takes advantage of parallelization, progressive computation and streaming, but assumes that the CFD data sets are computed in real-time.

3 The DSVR Framework

The *Distributed Simulation and Virtual Reality Environment* (DSVR) [6,8,9,11,12] is a 3D visualization framework, which implements all parts of the visualization pipeline as three networked instances (Fig. 1). These components are the 3D generator (data source, filter, mapper), the 3D streaming server, and the 3D viewer (rendering, presentation). The 3D generator consists of a software library (*libDVRP*) providing methods to be called directly from the compute nodes of the numerical simulation application. The data extraction and creation of 3D scenes, which represent features of the raw data, are efficiently implemented by parallel processing of the data parts based on the parallelization scheme of the respective application. The 3D scene sequences can be explored interactively by the 3D viewer.

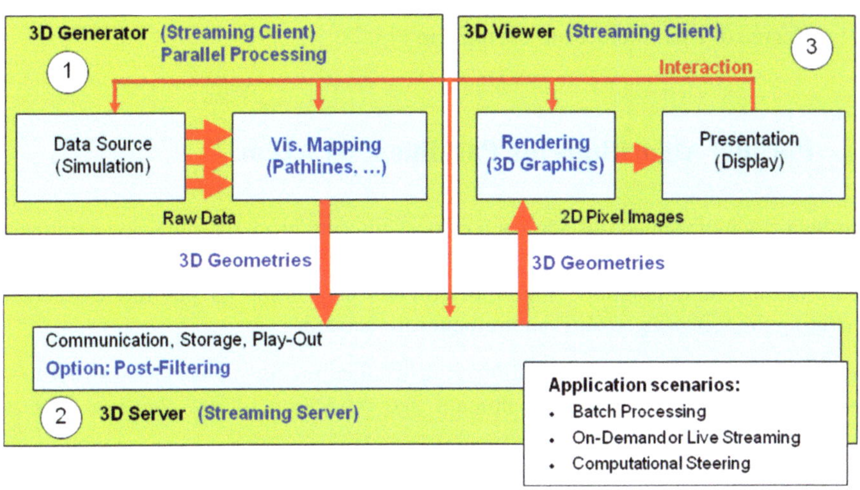

Fig. 1 Realization of the visualization pipeline in DSVR (www.dsvr-software.de)

The distribution of the visualization process chain leads to two advantages:

- The data volume to be stored is significantly reduced by storing the 3D geometries instead of raw data. This reduces the storage and bandwidth bottlenecks.
- Storing 3D geometries instead of rendered images supports highly interactive scenarios like 3D exploration of the scenes and immersive virtual reality.

One of the most time consuming parts of the visualization, the transmission of the calculated 3D scenes to the streaming server cannot be parallelized and tends to result in a bottleneck. Therefore the parallel speed-up of libDVRP can be improved by running it in split mode doing this task asynchronously on a separate cluster node (split node), hiding the corresponding latency [6].

To apply the concept of high-grade interaction without storing raw data to pathline extraction the initial line seeding becomes a special issue since numerical flow field analysis along more than one time step is not possible. On the other hand, storing an amount of homogeneously seeded pathlines will result in missing interesting features of the flow field when seeding not enough lines. However seeding too many lines leads to visual clutter in the form of a volume overfilled with lines where it will be nearly impossible to recognize any interesting feature. Additionally this would not be an approach supporting desired user interactions.

Recently libDVRP, 3D streaming server and viewer client were extended by a post-filtering approach [15], which still avoids the storing of raw data and also allows a high degree of interaction in real time and also in 3D video-on-demand scenarios. For this purpose a really huge amount of pathlines is seeded homogeneously all over the simulated volume. The 3D geometric representations of the pathlines are complemented by a set of properties at the current location and time. The property data can be calculated from the original vector field or given by the application. Afterwards a subset of pathlines can be selected in form of a "query over stream" by setting a query function during the playback of the 3D scenes. This not only results in a better understanding of the flow field features, but also allows a high degree of user interaction (example in Fig. 2).

4 Parallel Algorithms for Pathline Extraction

In Sect. 3 we presented our approach of post-filtering property-enhanced pathlines, which builds mainly on extraction of a huge amount of pathlines from the given raw data during the simulation. Therefore efficient algorithms for pathline extraction meeting the following *DSVR* requirements are needed:

- Support of massive parallelization on distributed memory systems using MPI.
- Support of the simulation's domain decomposition, where the raw data is available only locally.
- Support of on-the-fly processing of the data, because the data is available for one or two time steps only.

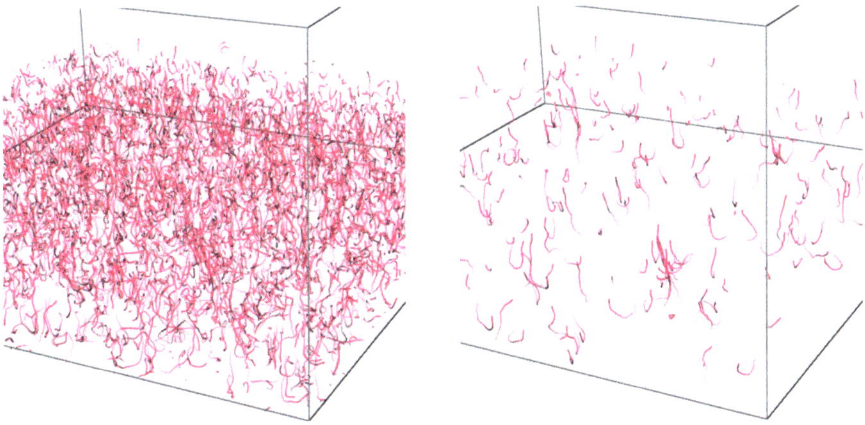

Fig. 2 Pathline visualization of a flow field in a simulation of atmospheric convection cells. Complete pathlines scene based on a generalized seeding strategy (*left*). Property-based post-filtered scene, exploring convection cells (*right*)

Table 1 Classification of parallel pathline extraction

	naive (4.2)	advanced (4.3)
data extraction	parallel	parallel
visualization mapping	parallel	asynchronous
3D geometry collecting	asynchronous	asynchronous

Especially the combination of the memory distribution and the domain decomposition prohibits the use of traditional approaches of dispersing the pathlines equally over the processes as it could be done on shared memory systems, where all threads have access to the complete raw data.

Below we will introduce two algorithms for parallel pathline extraction meeting these requirements (Table 1).

4.1 Seeding Strategies

Since there was a need for seeding many pathlines all over the vector field to support our approach of post-filtering property-enhanced pathlines there was a need for an adequate seed point placement, too. As mentioned above, it is not possible to calculate seed points by complex numerical functions without storing raw data. This is because it can not be determined which seed point would probably indicate an interesting feature in the future. For that reason the seed point placement can only be done with a generalized pattern with optional stochastic jitter.

The second challenge to meet for adequate seeding is the timing. Like it can not be determined where an interesting feature within the flow field occurs, it can not be prognosticated at which time this occurs. An additional problem arises when seeding lines only at the visualization's initialization: If the field has something like a main velocity direction due to fixed sources and drains, all lines will move into the drains while no lines remain at the sources. This is solved by seeding lines not once but in waves.

When seeding lines in waves it has to be assured that the total number of active lines does not exceed a certain limit. Therefore a line has a life time after which it disappears.

4.2 *Naive Parallelized Pathline Extraction*

Our first approach for parallel pathline extraction within the libDVRP library is based on the traditional approach of dispensing the lines over the compute nodes. However this dispensation does not happen stochastically with a view on load balancing but rather regarding to the simulation's domain decomposition. So each line with all its data is located at the process holding the needed part of the vector field.

Like all *DSVR* visualization the pathline extraction is done during the scientific simulation with the partial raw data in memory. During the initialization of the libDVRP each line is seeded within its belonging working node. At every time step of the simulation the libDVRP API is called in order to incrementally integrate all lines and to gather the resulting 3D geometries at the split node. Thereby on the working nodes the following is done (Fig. 3): Each node processes its lines by integrating them using Euler or Runge-Kutta techniques. This can lead to three issues:

1. The pathline still stays in the part of the grid volume, which is simulated in the actual process domain.
2. The pathline leaves the whole simulated grid volume and is not active any longer.
3. The pathline leaves the part of the grid volume of the actual process but still stays in the global grid.

In the third case the pathline is sent to the compute node holding the needed raw data. For performance optimization this send operations are buffered within a send queue. Using Runge-Kutta for integration, a pathline transfer could also be necessary during a Runge-Kutta sub-step. Since the Runge-Kutta integration of second order and above takes advantage of calculating intermediate vectors the pathline data structure is also capable of holding these vectors for transfer between the nodes.

Due to the fact that it is always indeterminable whether or not a process has to wait for more lines a synchronization feature is needed. So at the beginning of each visualization call all processes send their active line count to one master process. Furthermore each process informs this master process of finished lines. When all lines have finished, the master sends a termination message to all processes and the visualization routine is left.

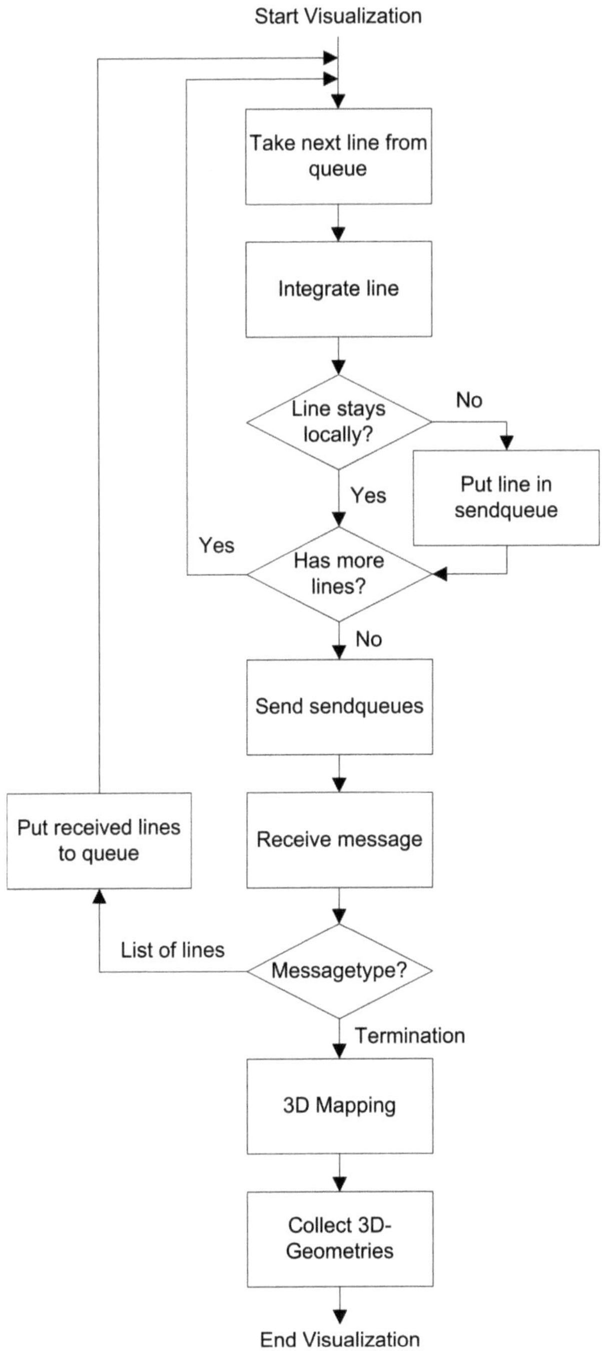

Fig. 3 Process chart for parallel in-element visualization

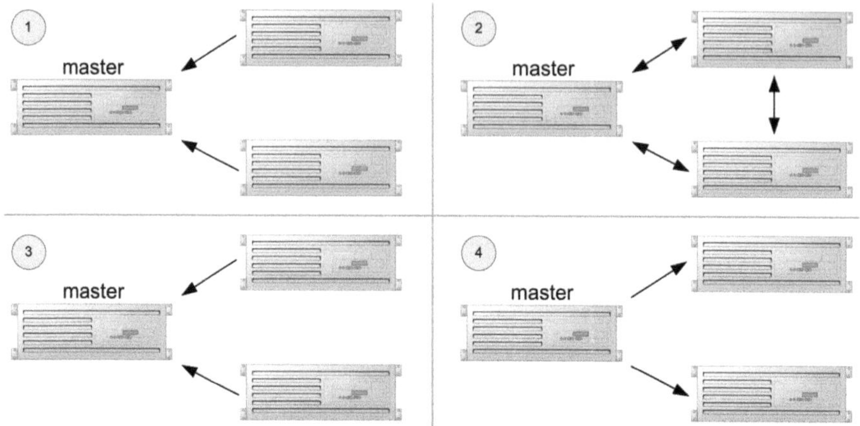

Fig. 4 MPI communication scheme: 1. gathering line counts on master process; 2. changing pathlines between processes holding neighbor domain data; 3. inform master process of finished lines; 4. termination sent by master process

This leads to the following four communication types (Fig. 4):

1. The active line count is gathered at the master process.
2. In case of pathlines leaving the local domain, these pathlines will be exchanged between processes. This includes the master process since it is part of the computing nodes holding parts of the vector field, too.
3. When one or more lines have finished, the count of locally finished lines is sent to the master process.
4. After all lines have finished, the master process broadcasts a termination message.

Besides these four communication types there is a need of a fifth communication for collecting the 3D geometries at the split node. This should not be neglected because of the transferred data volume, which is at least 20 Bytes per supporting point of every pathline. Assuming 30,000 pathlines with 100 supporting points each, this results in 57 MByte.

A series of tests has shown that even though the most parts of the algorithm are parallel it cannot really benefit from massive parallelization. The reason for poor speed-up is the big increase of the fraction of time which is spent for communication and synchronization.

4.3 Advanced Parallelized Pathline Extraction

Taking account of the gained knowledge from the first algorithm, we decided to go unusual ways for the integration of pathline extraction into the massively parallel libDVRP library: We started serializing parts of the parallel pathline extraction using the same concept well-proven for the streaming transfer in split mode.

This new concept uses two components (Fig. 5):

1. On the split node an additional thread for pathline extraction is started (pathline master).
2. The visualization routine on the compute nodes called by the simulation application.

All pathline data is located at the pathline master and the compute nodes do not know any lines. First the pathline master composes a list of points to integrate. For pathlines this would be the last supporting point of the line. For each point

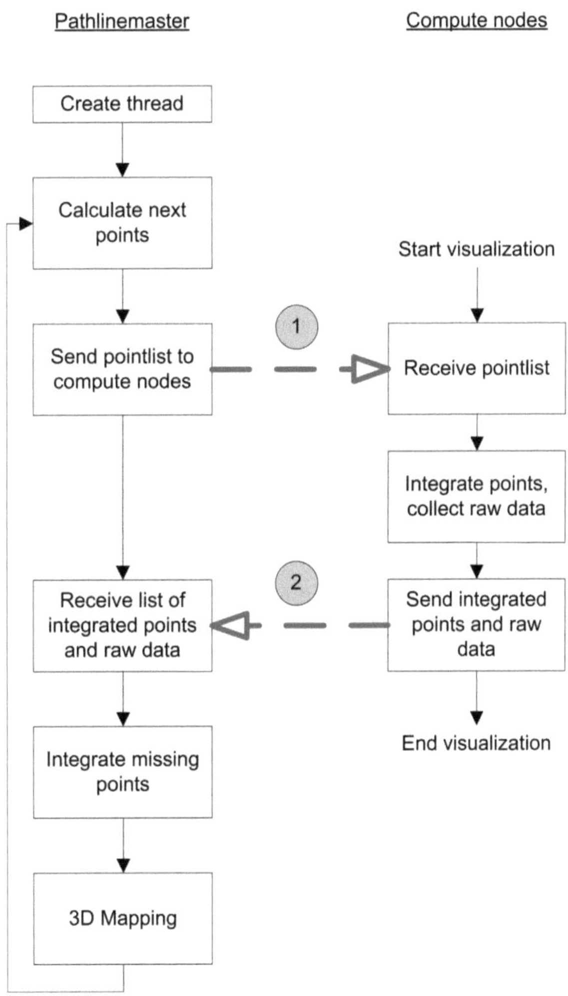

Fig. 5 Process chart for parallel in-element data extraction and subsequent concurrent 3D geometry mapping

to integrate the list also contains the indices of the grid cell containing the corresponding values. This list is sent to the compute nodes (Fig. 5 no. 1).

When the visualization routine on the compute nodes is called, the point list is received and separated into two parts, the points located in grid cells within the boundary layer and all other. For the first set, the raw data of the surrounding cells is collected while for the last set the compute node integrates the resulting points. Afterwards the nodes send the resulting data to the pathline master (Fig. 5 no. 2), the visualization routine ends and the compute nodes can go on with the simulation's next time step.

After receiving the integrated points and the raw data from the compute nodes, the pathline master starts integrating the points within the boundary layers. Subsequently the pathlines are converted into 3D geometries.

Like every time the split mode is used, the real network I/O is done by a separate write thread. But using an additional thread for the pathline master, the serial tasks of getting the list of points, integrating the points within the boundary layers and converting pathlines to 3D geometries is also done concurrently with the simulation application. This way these tasks are done asynchronously during the simulation.

For the presented algorithm there are some assets and drawbacks to note: On the one hand, since the algorithm uses fixed boundary layers, the maximum velocity of a grid cell may not lead in the particle exceeding the boundary layer. On the other hand, within the new algorithm there is no need for synchronization. It should also be mentioned that the data volume the split node collects could be significantly reduced. One further advantage of this algorithm is the flexible design regarding to the integration algorithm. Furthermore streaklines could be easily implemented by just modifying the point list generation.

4.4 A Combined Algorithm

Comparing the scaling results of the previous two algorithms, we could see the advantages of both algorithms. So the next issue was to combine them. The pathline extraction of the first algorithm was reduced to a simple particle tracing. This reduces the amount of data to be communicated between MPI processes when the lines' or particles' location changes between two of them. After the tracing is done, the resulting points are sent to the split node. The split node holds the pathlines and appends the new points to them. This also reduces the data volume to be collected. After that the visualization and storing of the pathlines is done.

5 Evaluation

For evaluation there was a need of finding an adequate simulation application meeting some key features. Purposing scalability tests the chosen application must show an adequate speed-up on its own. On the other hand, the test case should be a

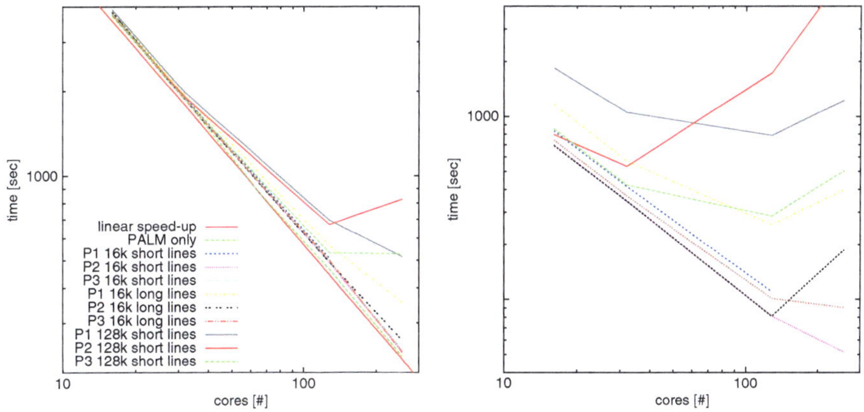

(a) time needed for the whole PALM run (b) time needed for the visualization task

Fig. 6 Scalability of the three algorithms P1 (Sect. 4.2), P2 (Sect. 4.3) and P3 (Sect. 4.4), different number of lines (16k, 128k) and supporting points (50, 200)

real simulation application instead of just an artificial data generator. As conclusion we use the PALM[1] code to simulate the airflow within a convection cell scenario.

This simulation has been done on a rectilinear grid with $320 * 320 * 80$ cells for 1,800 time steps. In each time step PALM calculates plenty of different data sets for each grid point. One of these data sets, a 3-dimensional vector containing the velocity of the air flow, was copied in a separated memory block and used by libDVRP for pathline extraction. Within this scenario we have measured the time consumed by the whole scenario as well as the time used for the visualization including the memory copy operations.

The testing environment was a part of the HLRN[2] cluster using nodes of the *ice1*. The *ice1* installation procures up to 320 "SGI Altix ICE 8200 Plus" nodes with 8 cores and 16 GB of memory on each node.

Within this test scenario we compared the three algorithms with each other. Additionally we tested the impact of the amount of seeded lines by seeding 16,000 and 128,000 lines, as well as the impact of the amount of the supporting points. Since for these tests we set our focus on the MPI communication and the parallel pathline extraction, we turned off the file writing part of the libDVRP library to get better measurement accuracy. Since the different algorithms use the same write mechanism this would not affect comparability. In the text below the core counts refer to the cores used for computing the simulation and don't include the split node. On the split node three cores are used in addition to these.

Looking on our first algorithm in Fig. 6b, it scales well up to 128 cores. Afterwards the visualization time increases slightly. This behavior could be observed

[1]PALM: *Parallelized Large-eddy Simulation Model*, Developed by Institute of Meteorology and Climatology, Leibniz University of Hannover.

[2]North-German Supercomputing Alliance (http://www.hlrn.de).

independently of the amount of lines or the amount of supporting points per line. This result was as expected as the extraction of pathlines does not take that much computation time and thereby the visualization time depends mainly on the MPI communication. Since the amount of data to be transferred by this algorithm mainly depends on the amount of lines and the amount of supporting points it does not surprise that visualizing more or longer pathlines will lead to longer run times. The rising visualization time with increasing core count is a tribute to the domain decomposition, which may lead to more or less lines to alternate processes. But even with this behavior we like to remember that the first algorithm does not need more than 0.3 seconds to visualize 128,000 pathlines on 256 cores.

Unlike the first algorithm, the second does not takes any advantage from parallelization but keeps consuming short visualization times until a certain degree of parallelization. After that the run time of the second algorithm increases dramatically. This is as expected as the algorithm consists of a very small parallel part and a larger serial part running concurrently to the simulation. At the point the simulation step gets running shorter than the serial pathline extraction, the algorithm starts blocking the simulation. Nevertheless applications can take advantage of this algorithm as long as a single simulation step need more time than a single serial pathline extraction step. This leads to the question for the time spent in the serial parts of the second algorithm. This time consists of three parts (see Fig. 5 in Sect. 4.3): The generation of the point list, the integration of points within the boundary layers and the conversion from pathline data into 3D geometries. The complexity of all three parts depends mainly on the count of lines. The complexity of the third part depends additionally on the lines' length.

The third algorithm scales symmetrically to the first algorithm at least until 256 cores used by the simulation. It also like the second algorithm consumes less run times for the visualization as the data volume of the collecting communication is reduced drastically. As in the second algorithm a situation where the visualization task on the split node will block the simulation nodes can be expected as the compute time the simulation will need for one time step gets less than the time needed for appending the particles to the pathlines on the split node.

Looking at Fig. 6b it gets clear, that all three introduced algorithms cannot take advantage of parallelization at all. So what does this mean for the scalability? To find out more about this, the run times of PALM with and without the different algorithms are compared in Fig. 6a. As expected PALM scales almost ideally when the visualization is switched off. In case of visualizing only a few lines with a restricted amount of supporting points all algorithms don't affect PALM scaling behavior that much. Visualizing longer pathlines the behavior of the second and third algorithm does not really change. On the other hand the first algorithm affects PALM noticeable as the data volume communicated for pathline collection would increase while the time needed for computation is not affected. Taking a look at the third test case visualizing 128k short pathlines all algorithms affect the total PALM runtime recognizable. Increasing the amount of lines to extract means increasing calculation time and increasing communication data volume on all three algorithms. It is also noticeable that until a certain degree of parallelization the visualization can take a benefit of an additional core dedicated to the pathline extraction.

6 Conclusion and Future Work

For numerical simulations where computational complexity and data volume prohibit the application of traditional explorative or post-processing visualization techniques for time-varying 3-dimensional vector fields, we have implemented a post-filtering approach which is based on data extracts consisting of densely seeded pathlines and properties at the individual current location. In order to contribute to the data extraction process, three highly scalable parallel algorithms for pathline extraction have been presented and evaluated on up to 256 cores.

For further work the serial part of the second and third algorithm should be reduced. This could be achieved by parallelizing this part in shared memory using Pthreads or OpenMP on a multi-core compute node. At the moment the pathlines are stored with the whole traces each time step. So in addition the data volume should be reduced by storing lines iteratively and by geometric compression.

Acknowledgement Computational support and infrastructure was provided by the North-German Supercomputing Alliance (HLRN). The authors are grateful to Prof. Dr. Siegfried Raasch and his group for sharing the PALM software and for fruitful discussions.

References

1. Bruckschen, R., Kuester, F., Hamann, B., Joy, K.I.: Real-time out-of-core visualization of particle traces. In: PVG '01: Proceedings of the IEEE 2001 symposium on parallel and large-data visualization and graphics, pp. 45–50 (2001)
2. Cheng, J.-R.C., Plassmann, P.E.: A parallel particle tracking framework for applications in scientific computing. J. Supercomput. **28**(2), 149–164 (2004)
3. Ellsworth, D., Green, B., Moran, P.: Interactive terascale particle visualization. In: VIS '04: Proceedings of the conference on Visualization '04, IEEE Computer Society, pp. 353–360. Washington, DC, USA (2004)
4. Gerndt, A., Hentschel, B., Wolter, M., Kuhlen, T., Bischof, C.: Viracocha: An efficient parallelization framework for large-scale cfd post-processing in virtual environments. In: SC '04: Proceedings of the 2004 ACM/IEEE conference on Supercomputing, p. 50, IEEE Computer Society, Washington, DC, USA (2004)
5. Al Globus: A Software Model for Visualization of Large Unsteady 3-D CFD Results. In: *AIAA 95-0115, 33rd Aerospace Sciences Meeting and Exhibit*, (1995)
6. Jensen, N., Olbrich, S., Pralle, H., Raasch, S.: An efficient system for collaboration in tele-immersive environments. In: Spencer, S.N. (Eds.) Proceedings of the Fourth Eurographics Workshop on Parallel Graphics and Visualization (EGPGV-02), pp. 123–132. 9–10 ACM Press. New York, September (2002)
7. Laramee, R., Hauser, H., Doleisch, H., Post, F., Vrolijk, B., Weiskopf, D.: The state of the art in flow visualization: Dense and texture-based techniques (2004)
8. Manten, S., Breuer, I., Olbrich, S.: Parallel isosurface extraction including polygon simplification via self adapting vertex clustering. In: The Ninth IASTED International Conference on Visualization, Imaging and Image Processing (VIIP), Cambridge, United Kingdom (2009)
9. Manten, S., Vetter, M., Olbrich, S.: Evaluation of a Scalable In-Situ Visualization System Approach in a Parallelized Computational Fluid Dynamics Application. In: Proceedings of Dagstuhl-Seminar "Virtual Realities", Schloss Dagstuhl 01.-06.06.2008, Springer (2011)

10. Mcloughlin, T., Laramee, R.S., Peikert, R., Post, F.H., Chen, M.: Over Two Decades of Integration-Based, Geometric Flow Visualization. In: Proceedings of Eurographics 2009
11. Olbrich, S., Manten, S., Jensen, N.: Scalable isosurface extraction in a parallelized streaming framework for interactive simulation and visualization. In: Proceedings of the 10th International Conference on Humans and Computers (HC-2007), pp. 147–152 (2007)
12. Olbrich, S., Pralle, H., Raasch, S.: Using streaming and parallelization techniques for 3D visualization in a high-performance computing and networking environment. In: *High-Performance Computing and Networking*, volume 2110 of *Lecture Notes in Computer Science*, pp. 231–240. Springer (2001)
13. Post, F., Laramee, R., Vrolijk, B., Hauser, H., Doleisch, H.: Feature extraction and visualisation of flow fields (2002)
14. Pugmire, D., Childs, H., Garth, C., Ahern, S., Weber, G.H.: Scalable computation of streamlines on very large datasets. In: SC '09: Proceedings of the Conference on High Performance Computing Networking, Storage and Analysis, New York (2009)
15. Vetter, M., Manten, S., Olbrich, S.: Exploring unsteady flows by parallel extraction of property-enhanced pathlines and interactive post-filtering. In: Proceedings of 14th Eurographics Symposium on Virtual Environments (EGVE 2007), Posters, pp. 9–12 (2008)
16. Yu, H., Ma, K.-L., Welling, J.: I/o strategies for parallel rendering of large time-varying volume data. In: Proceedings of the Eurographics/ACM SIGGRAPH Symposium on Parallel Graphics and Visualization, pp. 31–40, June (2004)
17. Yu, H., Wang, C., Ma, K.-L.: Parallel hierarchical visualization of large time-varying 3d vector fields. In: Proceedings of ACM/IEEE Supercomputing 2007 Conference, Reno, ACM/IEEE November (2007)

Towards Intelligent Management of Very Large Computing Systems

Eugen Volk, Jochen Buchholz, Stefan Wesner, Daniela Koudela, Matthias Schmidt, Niels Fallenbeck, Roland Schwarzkopf, Bernd Freisleben, Götz Isenmann, Jürgen Schwitalla, Marc Lohrer, Erich Focht, and Andreas Jeutter

Abstract The increasing complexity of current and future very large computing systems with a rapidly growing number of cores and nodes requires high human effort on administration and maintenance of these systems. Existing monitoring tools are neither scalable nor capable to reduce the overwhelming flow of information and provide only essential information of high value. Current management tools lack on scalability and capability to process a huge amount of information intelligently by relating several data and information from various sources together for making right decisions on error/fault handling. In order to solve these problems, we present a solution designed within the TIMaCS project, a hierarchical, scalable, policy based monitoring and management framework.

E. Volk (✉) · J. Buchholz · S. Wesner
High Performance Computing Center Stuttgart, Nobelstrasse 19, D-70569 Stuttgart, Germany
e-mail: volk@hlrs.de; buchholz@hlrs.de; wesner@hlrs.de

D. Koudela
Technische Universität Dresden, Zentrum für Informationsdienste und Hochleistungsrechnen (ZIH), D-01062 Dresden, Germany
e-mail: daniela.koudela@tu-dresden.de

M. Schmidt · N. Fallenbeck · R. Schwarzkopf · B. Freisleben
Department of Mathematics and Computer Science, University of Marburg, Hans-Meerwein-Str. 3, D-35032 Marburg, Germany
e-mail: schmidtm@informatik.uni-marburg.de; fallenbe@informatik.uni-marburg.de; rschwarzkopf@informatik.uni-marburg.de; freisleb@informatik.uni-marburg.de

G. Isenmann · J. Schwitalla · M. Lohrer
science + computing ag, Hagellocher Weg 73, D-72070 Tübingen, Germany
e-mail: isenmann@science-computing.de; j.schwitalla@science-computing.de; lohrer@science-computing.de

E. Focht · A. Jeutter
NEC High Performance Computing Europe,
Hessbruehlstrasse 21b, D-70565 Stuttgart, Germany
e-mail: efocht@hpce.nec.com; ajeutter@hpce.nec.com

C. Bischof (eds.), *Competence in High Performance Computing 2010*,
DOI 10.1007/978-3-642-24025-6_16,

1 Introduction

Operators of very large computing centres face the challenge of the increasing size of their offered systems following Moores or Amdahls law already for many years. Until recently the effort needed to operate such systems has not increased similarly thanks to advances in the overall system architecture, as systems could be kept quite homogeneous and the number of critical elements with comparably short Mean Time Between Failure (MTBF) such as hard disks could be kept low inside the compute node part.

Current petaflop and future exascale computing systems would require an unacceptable growing human effort for administration and maintenance based on an increased number of components. But even more would the effort rise due to their increased heterogeneity and complexity [1–3]. Computing systems cannot be built anymore with more or less homogeneous nodes that are similar siblings of each other in terms of hardware as well as software stack. Special purpose hardware and accelerators such as GPGPUs and FPGAs in different versions and generations, different memory sizes and even CPUs of different generations with different properties in terms of number of cores or memory bandwidth might be desirable in order to support not only simulations covering the full machine with a single application type, but also more coupled simulations exploiting the specific properties of a hardware system for different parts of the overall application. Different hardware versions go together with different versions and flavours of system software such as operating systems, MPI libraries, compilers, etc. as well as different, at best individual user specific, variants combining different modules and versions of available software fully adapted to the requirements of a single job. Additionally the operation model from purely batch might be complemented by usage models allowing more interactive or time controlled access for example for simulation steering or remote visualization jobs.

While the problem of detecting hardware failures such as a broken disk or memory has not changed and still can be done similarly as in the past by specific validation scripts and programs between two simulation jobs the problems that occur in relation with different software versions or only in specific use scenarios are much more complex to be detected and are clearly beyond what a human operator can address with a reasonable amount of time. Consequently the obvious answer is that the detection of problems based on different type of informations collected at different time steps needs to be automated and moved from the pure data level to the information layer where an analysis of the information either leads to recommendations to a human operator or at best trigger a process applying certain counter measures automatically.

A wide range of monitoring tools such as Ganglia [4] or ZenossCore [5] exist that are neither scalable to the system sizes of thousands of nodes and hundred thousands of compute cores, cannot cope with different or changing system configurations (e.g. this service is only available if the compute node is booted in certain OS modes), and the fusion of different informations to a consolidated system analysis state is missing, but more important they lack a powerful mechanism to analyse the informations monitored and to trigger reactions to change the system state actively to bring the system state back to normal operations.

Another major limitation is the lack of integration of historical data in the information processing, the lack of integration with other data sources (e.g. planned system maintenance schedule database) and the very limited amount of counter measures that can be applied. In order to solve these problems, we propose in scope of the TIMaCS [6] project a scalable, hierarchical policy based monitoring and management framework. The TIMaCS approach is based on an open architecture allowing the integration of any kind of monitoring solution and is designed to be extensible for information consumers and processing components. The design of TIMaCS follows concepts coming from the research domain of organic computing (e.g. see References [7] and [8]) also propagated by different computing vendors such as IBM in their autonomic computing [9] initiative.

In this paper we present the TIMaCS solution in form of a hierarchically structured monitoring and management framework, capable to solve the challenges and problems mentioned above.

2 Related Work

There are lots of tools available supporting the monitoring and management of large systems but they all originate from one of the two domains. Either the tools [10] (like Nagios [11], Ganglia [4], Zenoss [5]) are designed to monitor systems with only rudimentary management capabilities like executing a specific command for each failing sensor state. But they don't care about implications resulting from other failures. Additionally their scalability is limited in the focus of future high performance computing resources and i.e. current techniques visualizing the status will no longer be adequate due to the huge amount of data. Or the tools are designed to manage systems like Tivoli [12] which means to force to set up the machines according to an overall configuration which is normally done regardless of the underlying state. This is mostly done on a regular basis to force global changes down to all systems and even install needed software if not available and so on. But changing configurations in reaction to failing systems or services is not covered by them, so no real error handling can be done.

3 TIMaCS: Solution

The project TIMaCS (Tools for Intelligent System Management of Very Large Computing Systems) is initiated to solve the above mentioned issues. TIMaCS deals with the challenges in the administrative domain upcoming due to the increasing complexity of computing systems especially of computing resources with a performance of several petaflops. The project aims at reducing the complexity of the manual administration of computing systems by realizing a framework for intelligent management of even very large computing systems based on technologies for virtualization, knowledge-based analysis and validation of collected information, definition of metrics and policies.

The TIMaCS framework includes open interfaces which allow easy integration of existing or new monitoring tools, or binding to existing systems like accounting, SLA management or user management systems. Based on predefined rules and policies, this framework will be able to automatically start predefined actions to handle detected errors, additionally to the notification of an administrator. Beyond that the data analysis based on collected monitoring data, regression tests and intense regular checks aims at preventive actions prior to failures.

We seek for developing a framework ready for production and its validation at the High Performance Computing Center Stuttgart (HLRS), the Center for Information Services and High Performance Computing (ZIH) and the Distributed Systems Group at the Philipps University Marburg. NEC with the European High Performance Computing Technology Center and science + computing are the industrial partners within the TIMaCS project. The project funded by the German Federal Ministry of Education and Research started in January 2009 and will end in December 2011.

The following subsections describe the TIMaCS framework, presenting its architecture and components.

3.1 High Level Architecture

The description of the TIMaCS architecture provided in this section is based on earlier papers of the authors [13, 14]. In contrast to earlier papers, which described the TIMaCS architecture on a very high level, this paper presents the architecture in more detail, describing each component.

The self management concept of the proposed framework follows the IBM autonomic computing reference architecture [9]. Self managing of autonomic capabilities in computer systems perform tasks that IT professionals choose to delegate to the technology according to predefined policies and rules. Thereby, policies determine the type of decisions and actions that autonomic capabilities perform [9].

The TIMaCS framework is designed as a policy based monitoring and management framework with an open architecture and a hierarchical structure. The hierarchy of the TIMaCS framework is formed by management layers acting on different levels of information abstraction. This is achieved by generating state information for groups of different granularity: resource/node, node-group, cluster, organization. These granularities form abstraction layers. A possible realization of the hierarchies can be achieved in a tree-like structure, as shown in Fig. 1.

The bottom layer, called resource/node layer, contains resources or compute nodes with integrated sensors, which provide monitoring information about

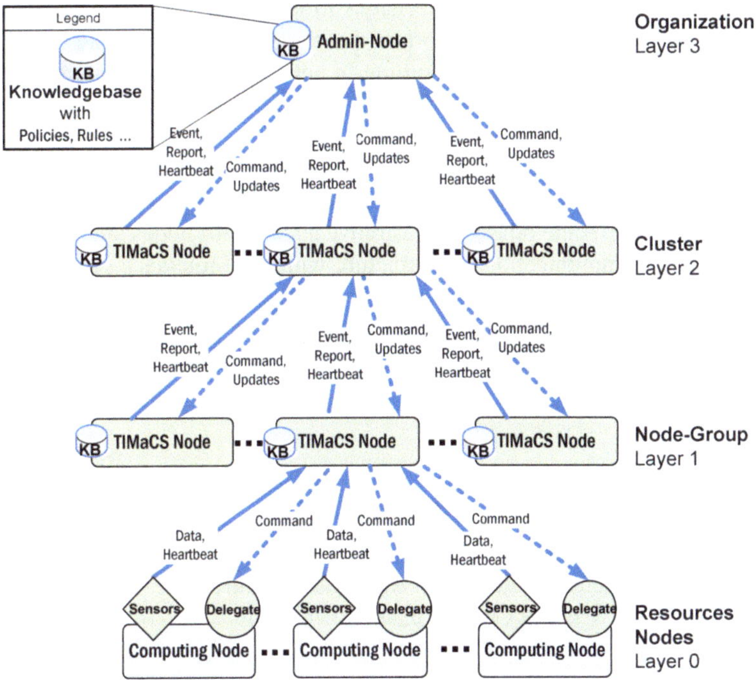

Fig. 1 The hierarchy of TIMaCS

resources or services running on them. Additionally, each managed resource has integrated Delegates. These are interfaces, which allow to execute commands on managed resources. Furthermore, there exist other Delegates which are not directly integrated in resources (e.g. Job Scheduler), but have indirect possibility to influence those resources (e.g. by removing error nodes from the batch queue).

Each management layer consists of dedicated nodes, called TIMaCS nodes, with monitoring and management capabilities organized in two logical blocks. The monitoring block collects information from the nodes of the underlying layer. It aggregates the information and makes conclusions by pre-analysing information, creating group states of certain granularity and triggering events, indicating possible errors. The management block analyses triggered events applying intelligent escalation strategies, and determines which of them need decisions. Decisions are made in accordance with predefined policies and rules, which are stored in a Knowledge Base filled up by system administrators when configuring the framework. Decisions result in commands, which are submitted to Delegates and executed on managed resources (compute nodes) or other components influencing managed resources, like a job-scheduler, capable to remove defective nodes from the batch queue. The hierarchic structure of the TIMaCS framework allows reacting on errors locally with very low latency. Each decision is reported to the upper layer to inform about detected error-events and selected decisions to handle the error-event. The upper layer, which generally has more information, is now able to intervene on received reports by making new decisions resulting in new commands, or an update of the knowledge base of the lower layer. Only escalated reports/events require effort of the administrator for deeper analysis.

On top of the framework an Admin-Node is settled, which allows administrators to configure the framework, the infrastructure monitoring, to maintain the knowledge base and to execute other administrative actions. All nodes are connected by message based communication infrastructure with fault tolerance capabilities and mechanisms ensuring the delivery of messages following the AMQP standard.

The following subsections describe the architecture of the TIMaCS components in detail, explaining monitoring, management and virtualisation of the system.

3.2 *Monitoring*

The monitoring capability of a TIMaCS node provided in the monitoring block, consists of Data-Collector, Storage, Aggregator, Regression Tests, Compliance Tests and the Filter & Event Generator, as shown in Fig. 2.

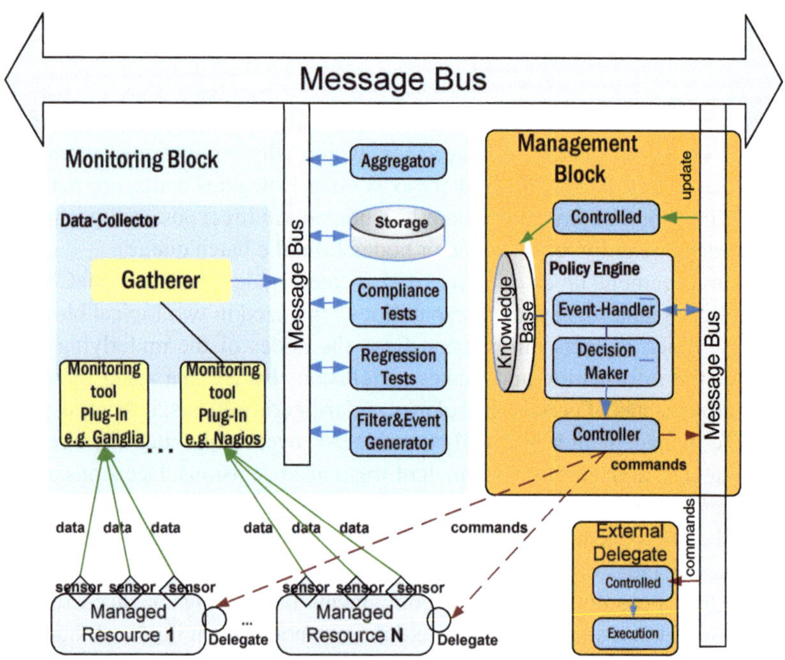

Fig. 2 Monitoring and Management in TIMaCS

The components within the monitoring block are connected by messaging middleware, enabling flexible publishing and consumption of data according to topics. These components are explained in the subsequent sections.

3.2.1 Data-Collector

The *Data-Collector* collects metric data and information about monitored infrastructure from different sources, including compute nodes, switches, sensors or other sources of information. The collection of monitoring data can be done synchronous or asynchronous, in pull or push manner, depending on the configuration of the component. In order to allow integration of various existing monitoring tools (like Ganglia [4] or Nagios[11]) or other external data-sources, we use a plug-in-based concept, which allows the design of customized plugins, capable to collect information from any data-source, as shown in Fig. 2. Collected monitoring data consist of metric values, and are semantically annotated with additional information, describing source location, the time, when the data were received, and other relevant information for data processing. Finally, the annotated monitoring data are published according to topics, using AMQP based messaging middleware, ready to be consumed and processed by other components.

3.2.2 Storage

The *Storage* subscribes to the topics published by the Data-Collector, and saves the monitoring data in the local round robin database. Stored monitored data can be retrieved by system administrators and by components analysing the history of the data, such as Aggregator or Regression Tests.

3.2.3 Aggregator

The *Aggregator* subscribes to topics produced by the data collector, and aggregates the monitoring data, i.e. by calculating average values, or the state of certain granularity (services, nodes, node-groups, cluster etc.). The aggregated information is published with new topics, to be consumed by other components of the same node (i.e. by the Filter & Event Generator), or those of the upper layer.

3.2.4 Regression Tests

The *Regression Tests* help cutting down on system outage periods by identifying components with a high probability of soon failure. Replacing those parts during regular maintenance intervals avoids system crashes and unplanned downtimes.

To get an indication, if the examined component may break in the near future, regression tests evaluate the chronological sequence of data for abnormal behaviour. The algorithm, which analyses those data, we call regression analysis. Since different metrics may need different algorithms for obtaining usable hints of the proper functioning of a component, TIMaCS allows for different regression analyses, which are implemented through an open interface.

One of the implemented regression analyses is the linear regression. There, a linear function is fitted to the data and the slope is returned. This algorithm is especially useful for predicting the state of a hard disk and evaluating memory errors on a DIMM.

TIMaCS distinguishes between online- and offline-regression tests. Online-regression tests are performed on a regular time interval and evaluate the most recent historical data being delivered by the publish/subscribe-system. Offline-regression tests on the contrary are only performed on request. They query the database to obtain their data for evaluation.

3.2.5 Compliance Tests

Compliance Tests enable early detection of software and/or hardware incompatibilities. They verify, if the correct versions of firmware, hardware and software are installed and they test, if every component is on the right place and working properly. Compliance tests are only performed on request since they are designed to run at the end of a maintenance interval or as a preprocessor to batch jobs. They may use the same sensors as used for monitoring but additionally they allow for starting bench marks.

Both compliance and regression tests are an integral part of TIMaCS. Therefore, they can easily be automated and help reducing the manual administrative costs.

3.2.6 Filter & Event Generator

The *Filter & Event Generator* subscribes to particular topics produced by the Data-Collector, Aggregators, and Regression- or Compliance Tests. It evaluates received data by comparing it with predefined values. In case that values exceed permissible ranges, it generates an event, indicating a potential error. The event is published according to a topic and sent to that components of the management block, which subscribed to that topic.

The evaluation of data is done according to predefined rules, defining permissible data ranges. These data ranges may differ depending on the location, where these events and messages are published. Furthermore, the possible kinds of messages and ways to treat them may vary strongly from site to site and in addition it depends on the layer the node belongs to.

The flexibility obviously needed can only be achieved by providing the possibility of explicitly formulating the rules by which all the messages are handled. TIMaCS provides a graphical interface for this purpose, based on eclipse Graphical Modelling Framework [15].

3.3 Management

The management block is responsible for making decisions in order to handle error events. It consists of the following components: Event Handler, Decision Maker, Knowledge Base, Controlled and Controller, as shown in Fig. 2. Subsequent sections describe these components in detail.

3.3.1 Event Handler

The *Event Handler* analyses received reports and events applying escalation strategies, to identify those which require error handling decisions. The analysis comprises methods evaluating the severity of events/reports and reducing the amount of related events/reports to a complex event. The evaluation of severity of events/reports is based on their frequency of occurrence and impact on health of affected granularity, as service, compute node, group of nodes, cluster etc. The identification of related events/reports is based on their spatial and temporal occurrence, predefined event relationship patterns, or models describing the topology of the system and dependencies between services, hardware and sensors. After the event has been classified as "requiring decision", it is handed over to the Decision Maker.

3.3.2 Decision Maker

The *Decision Maker* is responsible for planning and selecting error correcting actions, made in accordance with predefined policies and rules, stored in the Knowledge Base. The local decision is based on an integrated information view, reflected in a state of affected granularity (compute node, node group, etc.). Using the topology of the system and dependencies between granularities and subgranularities, the Decision Maker identifies the most probable origin of the error. Following predefined rules and policies, it selects decisions to handle identified errors. Selected decisions are mapped by the Controller to commands, and are submitted to nodes of the lower layer, or to Delegates of managed resources.

3.3.3 Knowledge Base

The *Knowledge Base* is filled up by the system administrators when configuring the framework. It contains policies and rules as well as information about the topology of the system and the infrastructure itself. Policies stored in the Knowledge Base are expressed by a set of objective statements prescribing the behaviour of the system on a high level, or by a set of (event, condition, action) rules defining actions to be executed in case of error detection, thus prescribing the behaviour of the system on the lower level.

3.3.4 Controller, Controlled and Delegate

The *Controller* component maps decisions to commands and submits these to *Controlled* components of the lower layers, or to *Delegates* of the managed resources.

The *Controlled* component receives commands or updates from the *Controller* of the management block of the upper layer and forwards these, after authentication and authorization, to addressed components. For example, received updates containing new rules or information, are forwarded to the Knowledge Base to update it.

The *Delegate* provides interfaces enabling the receipt and execution of commands on managed resources. It consists of *Controlled* and *Execution* components. The *Controlled* component receives commands or updates from the channels to which it is subscribed and maps these to device specific instructions, which are executed by the *Execution* component. In addition to *Delegates* which control managed resources directly, there are other *Delegates* which can influence the behaviour of the managed resource indirectly. For example, the virtualization management component, presented in Sect. 3.4, is capable to migrate VM-instances of affected or faulty nodes to healthy nodes.

3.4 Virtualization in TIMaCS

Virtualization is an important part of the TIMaCS project, since it enables partitioning of HPC resources. Partitioning means that the physical resources of the system are assigned to host and execute user-specific sets of virtual machines. Depending on the users' requirements, a physical machine can host one or more virtual machines that either use dedicated CPU cores or share the CPU cores. Virtual partitioning of HPC resources offers a number of benefits for the users as well as for the administrators. Users no longer rely on the administrators to get new software (including dependencies such as libraries) installed, but they can install all software components in their own virtual machine. Additional protection mechanisms including the virtualization hypervisor itself guarantee protection of the physical resources. Administrators benefit from the fact that virtual machines are easier to manage in certain circumstances than physical machines. One of the benefits of using TIMaCS is to have an automated system that makes decisions based on a complex set of rules. A prominent example is the failure of certain hardware components (e.g. fans) which leads to an emergency shutdown of the physical machines. Prior to the actual system shutdown, all virtual machines are live-migrated to another physical machine. This is one of the tasks of the TIMaCS virtualization component.

The used platform virtualization technology in the TIMaCS setup is the Xen Virtual Machine Monitor [16] since Xen with para-virtualization offers a reasonable tradeoff between performance and manageability. Nevertheless, the components

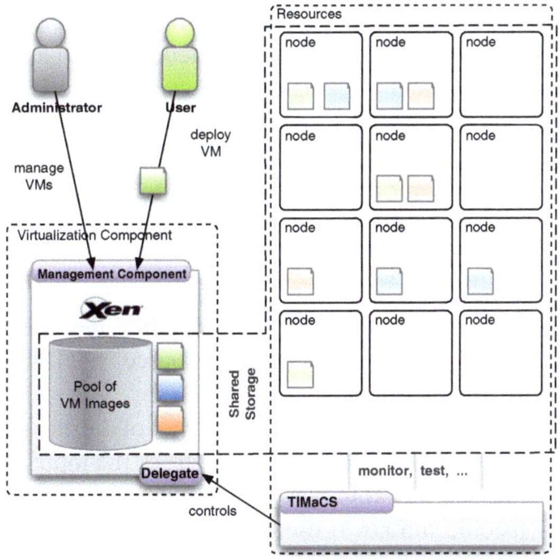

Fig. 3 TIMaCS Virtualization Components

are based on the popular *libvirt*[1] implementation and thus can be used with other hypervisors such as the Kernel Virtual Machine (KVM). The connection to the remaining TIMaCS framework is handled by a Delegate that receives commands and passes them to the actual virtualization component. A command could be the request to start a number of virtual machines on specific physical machines or the live migration from one machine to another. If the framework relies on a response, i.e. it is desirable to perform some commands synchronously, the Delegate responds back to an event channel.

Figure 3 describes the architecture of the TIMaCS virtualization components. The image pool plays a central rule since it contains all virtual machines' disk images either created by the user or the local administrator. Once a command is received via the Delegate, the virtualization component takes care of executing it.

3.5 Communication Infrastructure

To enable communication in TIMaCS, all TIMaCS nodes of the framework are connected by a scalable message based communication infrastructure supporting publish/subscribe messaging pattern, with fault tolerant capabilities and mechanisms ensuring delivery of messages, following the Advanced Message Queuing Protocol (AMQP) [17] standard. Communication between components of the same node is done internally, using memory based exchange channels bypassing the

[1] http://libvirt.org/.

communication server. In a topic-based publish/subscribe system, publishers send messages or events to a broker, identifying channels by unique URIs, consisting of topic-name and exchange-id. Subscribers use URIs to receive only messages with particular topics from a broker. Brokers can forward published messages to other brokers with subscribers that are subscribed to these topics.

The format of topics used in TIMaCS consists of several sub-keys (not all sub-keys need to be specified):

```
<source/target>.<kind>.<kind-specific>
```

- The sub-key *source/target* specifies the sender(group) or receiver(group) of the message, identifying a resource, a TIMaCS node or a group of message consumers/senders.
- The sub-key *kind* specifies the type of the message (data, event, command, report, heartbeat, ...), identifying a type of the topic consuming component.
- The sub-key *kind-specific* is specific to *kind*, i. e., for the kind "data", the kind-specific sub-key is used to specify the metric-name.

The configuration of the TIMaCS communication infrastructure comprises the setup of the TIMaCS nodes and AMQP based messaging middleware, connecting TIMaCS nodes according to the topology of the system. This topology is statically at the beginning of the system setup, but can be changed dynamically by system updates during run time. To build up a topology of the system, a connection between TIMaCS nodes and AMQP servers, the latter are usually colocated with TIMaCS nodes in order to achieve scalability, must follow a certain scheme. *Upstreams*, consisting of event-, hearbeat-, aggregated-metrics and report-messages, are published on messaging servers of the superordinated management node, enabling faster access to received messages. *Downstreams*, consisting of commands and configuration updates, are published on messaging servers of the local management node. This ensures that commands and updates are distributed in an efficient manner to addressed nodes or group of nodes.

Using an AMQP based publish/subscribe system, such as RabbitMQ [18], enables TIMaCS to build up a flexible, scalable and fault tolerant monitoring and management framework, with high interoperability and easy integration.

4 Conclusion

Challenges in the area of administration of very large computing systems have led to the design of the TIMaCS solution, a scalable policy based monitoring and management framework. From the system monitoring point of view, the presented TIMaCS framework reduces the overwhelming information flow of monitoring-data by handling and filtering it on different levels of abstractions. At the same time, it increases the information value delivered to the system administrator, comprising only necessary and important information. The usage of compliance- and regression tests enables administrators to realize preventive actions, allowing to check the status of the infrastructure prior to failures. The plug-in based monitoring concept enables integration of various existing monitoring tools like Ganglia, Nagios, ZenossCore and other information sources, so providers are not forced to replace their existing monitoring installation.

One big issue for system administrators, especially on HPC resources, is the capability to take actions in case of an error in a predefined way. The TIMaCS management framework supports different automisation and escalation strategies to handle errors based on policies, including notification of an administrator, semi-automatic to fully-automatic counteractions, prognoses, anomaly detection and their validation. Automated error handling reduces the system recovery time. The hierarchic structure of the framework allows reacting on errors locally with very low latency, although the full system status can be used for making a decision. The upper layers can intervene if their more global view leads to another decision.

The virtualization concept used in TIMaCS enables administrators easy partitioning and dynamic user assignment of very large computing systems, allowing setup, migration or removal of single compute nodes out of a heterogeneous or hybrid system.

Using the AMQP based publish/subscribe system enables TIMaCS to build up a flexible, scalable and fault tolerant monitoring and management framework, with high interoperability and easy integration.

By installing the TIMaCS framework the administrator will be enabled to specify rules and policies with simple access to all monitored data. So it is not necessary to know any details about the underlying monitoring systems since the sensor information is standardized. So defining error handlings becomes very simple and they can also be activated in different manners from manually on demand to fully automated when the actions are well tested. Going far beyond current practice it is even possible to define lots of different cluster configurations and set them up in different partitions of the whole system in parallel or changing over time. So it is possible to cut off a specific part of the cluster for urgent computing with higher storage bandwidth for some time or change the scheduling according to the changing submit behaviour from weekdays to weekends. It might even be possible with some further developments to allow users to define which hardware characteristics they need i.e. minimal network but high compute power may result in a restriction of the network so that the overall network performance can be assured to other users. This may result in more detailed payment models.

Thus TIMaCS will give the administrator a tool so that he can manage increasing system complexity and handle current issues like errors and even be prepared to do changes very dynamically, which currently need very long time consuming actions with a lot of manual overhead, so that in practice it is done only once when installing the cluster or extending it.

Acknowledgements The results presented in this paper are partially funded by Federal Ministry of Education and Research (BMBF) through the TIMaCS [6] project.

References

1. Strohmaier, E., Dongarra, J.J., Meuer, H.W., Simon, H.D.: Recent trends in the marketplace of high performance computing, Parallel Computing, Volume 31, Issues 3–4, pp. 261–273 March-April (2005)
2. Wong, Y.W., Mong Goh R.S., Kuo, S., Hean Low, M.Y.: A Tabu Search for the Heterogeneous DAG Scheduling Problem, 15th International Conference on Parallel and Distributed Systems (2009)
3. Asanovic, K., Bodik, R., Demmel, J., Keaveny, T., Keutzer, K., Kubiatowicz, J., Morgan, N., Patterson, D., Sen, K., Wawrzynek, J., Wessel, D., Yelick, K.: A view of the parallel computing landscape, Communications of the ACM, v. 52 n. 10, October (2009)
4. Ganglia web-site, http://ganglia.sourceforge.net/
5. Zenoss web-site, http://www.zenoss.com
6. TIMaCS project web-site, http://www.timacs.de
7. organic-computing web-site, http://www.organic-computing.de/spp
8. Wuertz, R.P.: Organic Computing (Understanding Complex Systems), Springer 2008
9. IBM: An architectural blueprint for autonomic computing, http://www-03.ibm.com/autonomic/pdfs/AC_Blueprint_White_Paper_V7.pdf, IBM Whitepaper, June 2006. Cited 16 December 2010
10. Linux Magazin, Technical Review, Monitoring, 2007
11. Nagios web-site, http://www.nagios.org
12. IBM Tivoli web-site, http://www-01.ibm.com/software/tivoli/
13. Buchholz, J., Volk, E.: The Need for New Monitoring and Management Technologies in Large Scale Computing Systems. In: Procedings of eChallenges 2010, to appear
14. Buchholz, J., Volk, E.: Towards an Architecture for Management of Very Large Computing Systems. In: Resch, M., Benkert, K., Wang, X., Galle, M., Bez, W., Kobayashi, H., Roller, S. (eds.) High Performance Computing on Vector Systems 2010, Springer, Berlin (2010)
15. eclipse Graphical Modeling Project (GMP) http://www.eclipse.org/modeling/gmp/
16. Barham, P., Dragovic, B., Fraser, K., Hand, S., Harris, T., Ho, A., Neugebauer, R., Pratt, I., Warfield, A.: Xen and the Art of Virtualization, in SOSP '03: Proceedings of the 19th ACM Symposium on Operating Systems Principles, ACM Press, Bolton Landing, NY, USA (2003)
17. Advanced Message Queuing Protocol (AMQP) web-site, http://www.amqp.org
18. RabbitMQ web-site, http://www.rabbitmq.com

Parallel Multigrid Methods and Coarse Grid LDL^T Solver for Maxwell's Eigenvalue Problem

Daniel Maurer and Christian Wieners

Abstract We consider efficient numerical solution methods for Maxwell's eigenvalue problem in bounded domains. A suitable finite element discretization with Nédélec elements on tetrahedra leads to large linear systems of equations, where the resulting matrices are symmetric but singular and where the kernel corresponds to divergence-free vector fields. The discretized eigenvalue problem is solved by a preconditioned iterative eigenvalue solver, using a modification of the LOBPCG method extended by a projection onto the divergence-free vector fields, where we use a multigrid preconditioner for a regularized Maxwell problem as well as for the solution of the projection problem. In both cases, a new parallel direct block LDL^T decomposition is used for the solution of the coarse grid problem.

Keywords Parallel computing • finite elements • block LDL^T decomposition • Maxwell's eigenvalue problem • LOBPCG method • multigrid preconditioner

1 Introduction

The propagation of electro-magnetic waves is described by the system of Maxwell's equations: determine the magnetic field $\mathbf{H}$ and the electric field $\mathbf{E}$ such that

$$\nabla \times \mathbf{H} - \varepsilon \partial_t \mathbf{E} = \mathbf{J} \qquad \nabla \times \mathbf{E} + \mu \partial_t \mathbf{H} = 0 \tag{1a}$$

$$\nabla \cdot (\varepsilon \mathbf{E}) = \rho \qquad \nabla \cdot (\mu \mathbf{H}) = 0 \tag{1b}$$

D. Maurer · C. Wieners (✉)
Institute for Applied and Numerical Mathematics 3, Karlsruhe Institute of Technology, 76128 Karlsruhe, Germany
e-mail: daniel.maurer@kit.edu; christian.wieners@kit.edu

C. Bischof (eds.), *Competence in High Performance Computing 2010*,
DOI 10.1007/978-3-642-24025-6_17,

hold in a medium described by the permittivity $\varepsilon = \varepsilon_0\varepsilon_\mathrm{r}$ and permeability $\mu = \mu_0\mu_\mathrm{r}$, and where $\mathbf{J}$ is the electric current density, and ρ is the electric charge density. The speed of light is $c = (\varepsilon_0\mu_0)^{-1/2}$.

For $\mathbf{J} = \mathbf{0}$, $\rho = 0$ (no external sources), the Maxwell system can be reduced to a second order equation for $\mathbf{E}$ or $\mathbf{H}$, respectively:

$$\nabla \times \left(\frac{1}{\varepsilon}\nabla \times \mathbf{H}\right) = -\mu\partial_t^2\mathbf{H}\,, \qquad \nabla \times \left(\frac{1}{\mu}\nabla \times \mathbf{E}\right) = -\varepsilon\partial_t^2\mathbf{E} \tag{2}$$

(it is sufficient to determine only one of the fields $\mathbf{E}$ or $\mathbf{H}$).

We consider only time-harmonic solutions of the form $\mathbf{u}(x,t) = \exp(\mathrm{i}\omega t)\mathbf{u}(x)$. For $\mathbf{u} = \mathbf{H}$ and the case $\mu_\mathrm{r} \equiv 1$ (no magnetic induction) this leads to

$$\nabla \times \left(\varepsilon_\mathrm{r}^{-1}\nabla \times \mathbf{u}\right) - (\omega/c)^2\mathbf{u} = 0\,. \tag{3}$$

This is the Maxwell eigenvalue problem, and the solution determines the resonating frequencies $\omega \neq 0$ of monochromatic waves in this medium. Here, we discuss efficient parallel finite element solution methods to compute these frequencies in bounded domains. Therefore, we use standard curl-conforming finite elements, a preconditioned Krylov space iteration for the approximation of eigenfunctions, and multigrid preconditioner. Note that the eigenspace for $\omega = 0$ are all divergence-free vector fields, so that the Krylov method has to be extended by a projection onto its orthogonal complement.

In particular we consider configurations where the coarse mesh in the multigrid method has more than 10,000 cells for an appropriate geometry resolution. Since the preconditioner for the Krylov space method is nearly singular in our application, we use a direct solver for the coarse problem. For this purpose, we introduce a new parallel LDL^T decomposition method which is based on a nested dissection approach together with the distributed parallel solution of the dense Schur complements.

The paper is organized as follows. We introduce the discrete Maxwell problem, its parallel finite element formulation and the eigenvalue solver in Section 2. The direct solver and the multigrid method is presented in Section 3. Finally, two representative examples demonstrate the performance of our methods.

2 The Discrete Maxwell Eigenvalue Problem

Let $\Omega \subset \mathbb{R}^3$ be a bounded polygonal domain, let

$$\mathbf{V}_h \subset H_0(\mathrm{curl},\Omega) = \{\mathbf{u} \in L_2(\Omega,\mathbb{R}^3)\colon \nabla \times \mathbf{u} \in L_2(\Omega,\mathbb{R}^3),\ \mathbf{u}\times\mathbf{n} = 0 \text{ on } \partial\Omega\} \tag{4}$$

be lowest order Nédélec elements on a tetrahedral mesh [5, Ch. 5], and let $Q_h \subset H_0^1(\Omega)$ be linear Lagrange elements. Then, the discrete eigenvalue problem in weak form reads as follows: find eigenfunctions $\mathbf{u}_h \in \mathbf{V}_h \setminus \{\mathbf{0}\}$ and eigenvalues $\lambda_h \in \mathbb{R}$ such that

$$(\varepsilon_{\mathrm{r}}^{-1}\nabla \times \mathbf{u}_h, \nabla \times \mathbf{v}_h)_0 = \lambda_h\,(\mathbf{u}_h, \mathbf{v}_h)_0\,, \qquad \mathbf{v}_h \in \mathbf{V}_h\,, \tag{5a}$$

$$(\mathbf{u}_h, \nabla q_h)_0 = 0\,, \qquad q_h \in Q_h\,. \tag{5b}$$

Since the curl-bilinear form is symmetric and elliptic in $\mathbf{X}_h = \{\mathbf{v}_h \in \mathbf{V}_h : \operatorname{div} \mathbf{v}_h = 0\}$, all eigenvalues are positive. Let A_h, M_h, B_h, and C_h be discrete operators defined by

$$\langle A_h\mathbf{u}_h, \mathbf{v}_h\rangle = (\varepsilon^{-1}\operatorname{curl}\,\mathbf{u}_h, \operatorname{curl}\,\mathbf{v}_h)_0\,, \qquad \langle M_h\mathbf{u}_h, \mathbf{v}_h\rangle = (\mathbf{u}_h, \mathbf{v}_h)_0\,, \tag{6a}$$

$$\langle B_h\mathbf{u}_h, q_h\rangle = (\mathbf{u}_h, \nabla q_h)_0\,, \qquad \langle C_h p_h, q_h\rangle = (\nabla p_h, \nabla q_h)_0 \tag{6b}$$

for $\mathbf{u}_h, \mathbf{v}_h \in \mathbf{V}_h$ and $p_h, q_h \in Q_h$. This rewrites the eigenvalue problem as

$$A_h\mathbf{u}_h = \lambda_h\, M_h\mathbf{u}_h\,, \qquad B_h\mathbf{u}_h = 0\,. \tag{7}$$

For the solution of the discrete Maxwell eigenvalue problem, we use a modified LOBPCG method introduced by Knyazev [3]. This method combines a three-term recursion with a preconditioned projected inverse iteration as proposed in [10], see Algorithm 1. The Krylov space for the iterative eigensolver is spanned by the projection $P_h = \mathrm{id} - B_h^T C_h^{-1} B_h : \mathbf{V}_h \longrightarrow \mathbf{X}_h$ of $\operatorname{span}\{(T_h^{-1}A_h)^j\mathbf{v}_h^0 : j = 0, \ldots, J\}$, where $T_h : \mathbf{V}_h' \longrightarrow \mathbf{V}_h$ is an approximate inverse of $A_h^\delta = A_h + \delta M_h$ (a regularization of the singular operator A_h for a small parameter $\delta > 0$). The (approximate) evaluation of the projection requires the sufficiently close evaluation of C_h^{-1} in steps S0) and S4).

3 A Parallel Block LDL^T Decomposition

In principle, for the coarse problem general parallel direct solver for sparse linear systems such as MUMPS [1] or PARDISO [6, 7] can be used. Here, it turns out that it is more efficient to use directly the symmetry and the parallel data structure of the finite element matrices. Therefore, we introduce a new LDL^T block decomposition with parallel Schur complement computation.

We shortly summarize the notation (for more details see [4], where the more general case of a parallel block LU decomposition is discussed). We begin with a general setting for finite element problems. Using the parallel programming model introduced in [8, 9], we define a parallel additive representation of the stiffness matrix which directly corresponds to a parallel domain decomposition.

Let $\mathscr{P} = \{1, \ldots, P\}$ be the set of processors (for simplicity let $P = 2^S$), and let

$$\overline{\Omega} = \bigcup_{p\in\mathscr{P}} \overline{\Omega}^p \tag{8}$$

Algorithm 1 The modified LOBPCG method (including projections)

S0) Choose randomly $\mathbf{v}_h^1, \dots, \mathbf{v}_h^N \in \mathbf{V}_h$. Compute $\mathbf{w}_h^n = P_h \mathbf{v}_h^n \in \mathbf{X}_h$ and normalize them.

S1) Ritz-step: Set up Hermitian matrices

$$\hat{A} = \Big(a(\mathbf{w}_h^m, \mathbf{w}_h^n)\Big)_{m,n=1,\dots,N}, \quad \hat{M} = \Big(m(\mathbf{w}_h^m, \mathbf{w}_h^n)\Big)_{m,n=1,\dots,N} \in \mathbb{C}^{N\times N}$$

and solve the matrix eigenvalue problem $\hat{A}\hat{z}^n = \lambda^n \hat{M}\hat{z}^n$.

S2) Compute $\mathbf{y}_h^n = \sum_{n=1}^{N} \hat{z}_m^n \mathbf{w}_h^n \in \mathbf{X}_h$.

S3) Compute $\mathbf{r}_h^n = A_h \mathbf{y}_h^n - \lambda^n M_h \mathbf{y}_h^n \in \mathbf{V}_h'$, check for convergence.

S4) Compute $\mathbf{v}_h^n := T_h \mathbf{r}_h^n \in \mathbf{V}_h$ and $\mathbf{w}_h^n = P_h \mathbf{u}_h \in \mathbf{X}_h$ for $n = 1, \dots, N$.

S5) Perform this step in the first iteration, otherwise go to step S6).
Perform Ritz-step for $\{\mathbf{w}_h^1, \dots, \mathbf{w}_h^N, \mathbf{v}_h^1, \dots, \mathbf{v}_h^N\} \subset \mathbf{X}_h$ of size $2N$ and set for $n = 1, \dots, N$

$$\mathbf{d}_h^n = \sum_{l=1}^{N} z_l^n \mathbf{w}_h^l \in \mathbf{X}_h, \qquad \tilde{\mathbf{y}}_h^n = \mathbf{d}_h^n + \sum_{l=1}^{N} z_{N+l}^n \mathbf{y}_h^l \in \mathbf{X}_h.$$

Set $\mathbf{y}_h^n = \tilde{\mathbf{y}}_h^n$ for $n = 1, \dots, N$. Go to step S2).

S6) Perform Ritz-step for $\{\mathbf{w}_h^1, \dots, \mathbf{w}_h^N, \mathbf{y}_h^1, \dots, \mathbf{y}_h^N, \mathbf{d}_h^1, \dots, \mathbf{d}_h^N\}$ of size $3N$ and set for $n = 1, \dots, N$

$$\tilde{\mathbf{d}}_h^n = \sum_{l=1}^{N} z_l^n \mathbf{w}_h^l + z_{2N+l}^n \mathbf{d}_h^l, \qquad \tilde{\mathbf{y}}_h^n = \mathbf{d}_h^n + \sum_{l=1}^{N} z_{N+l}^n \mathbf{y}_h^l.$$

Set $\mathbf{y}_h^n = \tilde{\mathbf{y}}_h^n$ and $\mathbf{d}_h^n = \tilde{\mathbf{d}}_h^n$ for $n = 1, \dots, N$. Go to step S2).

---

be a non-overlapping domain decomposition. Let $S_h = \operatorname{span}\{\phi_i : i \in \mathscr{I}\}$ be a finite element space spanned by nodal functions ϕ_i corresponding to nodal points $\mathbf{z}_i \in \bar{\Omega}$, and where $\mathscr{I}$ is a finite index set. The domain decomposition induces an overlapping decomposition $\mathscr{I} = \bigcup_{p\in\mathscr{P}} \mathscr{I}^p$ with $\mathscr{I}^p = \{i \in \mathscr{I} : z_i \in \overline{\Omega}^p\}$. For $i \in \mathscr{I}$ we define

$$\pi(i) = \big\{p \in \mathscr{P} : i \in \mathscr{I}^p\big\}. \tag{9}$$

Let $\Pi = 2^{\mathscr{P}}$ be the set of all possible processor sets. We define the subset of all active processor sets $\Pi_{\mathscr{I}} = \{\pi(i) \in \Pi : i \in \mathscr{I}\}$ with respect to the corresponding matrix graph. Next, we introduce a numbering on $\Pi_{\mathscr{I}}$. Depending on a processor numbering $\mathscr{P} = \{p_1, p_2, ..., p_{2^S}\}$ we define for each step $s = 0, \dots, S$ a cluster of combined processor sets $\mathscr{P}^{s,t} = \{p_j : 2^s(t-1) < j \le 2^s t\}$ for $t = 1, \dots, T_s = 2^{S-s}$. For every $\pi \in \Pi_{\mathscr{I}}$ we define the associated step by

$$s(\pi) = \min\left\{s \mid \exists t \in \{1, \dots, T_s\} : \pi \subset \mathscr{P}^{s,t}\right\}, \tag{10}$$

which results in

$$\Pi^s = \{\pi \in \Pi_{\mathscr{I}} : s(\pi) = s\} \qquad \Pi^{s,t} = \{\pi \in \Pi^s : \pi \subset \mathscr{P}^{s,t}\}. \tag{11}$$

Now, we can define a consecutive numbering for the disjoint decomposition

$$\Pi_{\mathscr{I}} = \{\pi_1, \pi_2, \dots, \pi_K\} = \Pi^0 \cup \dots \cup \Pi^S = \bigcup_{s=0}^{S} \bigcup_{t=1}^{T_s} \Pi^{s,t}, \tag{12}$$

i.e. $\Pi^{s,t} = \{\pi_k : k \in \mathscr{K}^{s,t}\}$ with $\mathscr{K}^{s,t} = \{K_{s,t-1} + 1, \dots, K_{s,t}\}$, $K_{0,0} = 0$, $K_{s,T_s} = K_{s+1,0}$, and $K = K_{S,1}$.

Let $a(\cdot,\cdot) = \sum_{p \in \mathscr{P}} a_p(\cdot,\cdot)$ be a symmetric bilinear form with additive contributions on Ω^p. This defines the entries of the stiffness matrix A in additive form:

$$A[i,j] = \sum_{p \in \mathscr{P}} A^p[i,j], \qquad A^p[i,j] = a_p(\phi_i, \phi_j). \tag{13}$$

For the parallel LDL^T decomposition we use the block matrices

$$A^p_{km} = (A^p[i,j])_{i \in \mathscr{I}_k, j \in \mathscr{I}_m} \in \mathbb{R}^{N_k \times N_m} \text{ with } \mathscr{I}_k = \{i \in \mathscr{I} : \pi(i) = \pi_k\}. \tag{14}$$

It can be shown that the sparsity pattern induced by the finite element assembling together with the numbering allows for a reduced LDL^T decomposition, cf. Algorithm 2 [4, Lem. 3]. For the diagonal blocks we use LAPACK routines for the LDL^T decomposition of $A_{kk} := \mathrm{LDL}(A_{kk})$ in line 4, and this decomposition is used for the computation of $A_{kk}^{-1}A_{km} := \textsc{Solve}(A_{kk}, A_{km})$ in line 8.

The parallel block LDL^T decomposition uses a (cyclic) distribution of the columns of the block matrices A_{kn} for $k, n \in \mathscr{K}^{s,t}$. For this purpose we select a processor map

$$p^{s,t} : \mathscr{K}^{s,t} \to \mathscr{P}^{s,t}, \tag{15}$$

and for $n \in \mathscr{K}^{s,t}$ the block matrices A_{kn}, $k \in \mathscr{K}^{s,t}$, are computed on processor $p = p^{s,t}(n)$. This results in the following parallel realization of Algorithm 2 by Algorithm 3, where all required parallel communication (i.e., the receive and send commands) are completely determined by the mapping $p^{s,t}$.

Algorithm 2 Reduced block LDL^T decomposition for finite element matrices

FOR $s = 0, \dots, S$
  FOR $t = 1, \dots, T_s$
    FOR $k = K_{s,t-1} + 1, \dots, K_{s,t}$
      $A_{kk} := \mathrm{LDL}(A_{kk})$
      FOR $n \in \mathcal{K}^{s,t}, n > k$
        $B_{kn} := A_{kn}$
      FOR $m \in \mathcal{K}^{s,t}, m > k$
        $A_{km} := \textsc{Solve}(A_{kk}, A_{km})$
        FOR $n \in \mathcal{K}^{s,t}, m \geq n > k$
          $A_{nm} := A_{nm} - B_{kn}^T A_{km}$

Algorithm 3 Parallel block LDL^T decomposition

FOR $s = 0, \dots, S$
  FOR $t = 1, \dots, T_s$
    IF $(s > 0)$
      FOR $n, m \in \mathcal{K}^{s,t}, m \geq n$
        $p := p^{s,t}(m)$
        ON $q := p^{s-1,2t}(m)$: SEND A_{nm}^q TO p
        ON $r := p^{s-1,2t+1}(m)$: SEND A_{nm}^r TO p
        ON p: RECEIVE $A_{nm}^{(q)}$ FROM q AND $A_{nm}^{(r)}$ FROM r
          AND SET $A_{nm}^p := A_{nm}^{(q)} + A_{nm}^{(r)}$
    FOR $k = K_{s,t-1} + 1, \dots, K_{s,t}$
      FOR $n \in \mathcal{K}^{s,t}, n > k$
        $p := p^{s,t}(n)$
        ON p: SEND A_{kn}^p TO $q \in \mathcal{P}^{s,t}$
        ON $q \in \mathcal{P}^{s,t}$: RECEIVE $A_{kn}^{(p)}$ AND SET $B_{kn}^q := A_{kn}^{(p)}$
      $p := p^{s,t}(k)$
      ON p: $A_{kk}^p := \mathrm{LDL}(A_{kk}^p)$
      ON p: SEND A_{kk}^p TO $q \in \mathcal{P}^{s,t}$
      ON $q \in \mathcal{P}^{s,t}$: RECEIVE $A_{kk}^{(p)}$ AND SET $B_{kk}^q := A_{kk}^{(p)}$
      FOR $m \in \mathcal{K}^{s,t}, m > k$
        $q := p^{s,t}(m)$
        ON q: $A_{km}^q := \textsc{Solve}(B_{kk}^q, A_{km}^q)$
        FOR $n \in \mathcal{K}^{s,t}, m \geq n > k$
          ON q: $A_{nm}^q := A_{nm}^q - (B_{kn}^q)^T A_{km}^q$

Now, we consider a mesh hierarchy and corresponding nested sequences of curl-conforming finite element spaces $\mathbf{V}_0 \subset \mathbf{V}_1 \subset \cdots \subset \mathbf{V}_J$ and H^1-conforming spaces $Q_0 \subset Q_1 \subset \cdots \subset Q_J$.

Algorithm 4 Multigrid preconditioner $v_J = T_J r_J$ with hybrid smoother.

S0) Set $j = J$.

S1) Set $v_j = 0$. For $\kappa = 1, \dots, \mu$ compute

$$w_j = R_j r_j\,, \qquad v_j := v_j + w_j\,, \qquad r_j := r_j - A_j^\delta w_j\,.$$

S2) For $\kappa = 1, \dots, \nu$ compute

$$w_j = \delta^{-1} A_j^\delta S_j D_{j-1} S_j' r_j\,, \qquad v_j := v_j + w_j\,, \qquad r_j := r_j - A_j^\delta w_j\,.$$

S3) If $j > 0$, set $r_{j-1} = I_j' r_j$, $j := j - 1$, and go to S1.

S4) If $j = 0$, set $v_0 := T_0 r_0$.

S5) For $j = 1, \dots, J$ set $v_j := v_j + I_j v_{j-1}$ and return the result v_J.

The parallel direct solver is used on the coarse level directly as preconditioner T_0 and for the projection P_0 in the LOBPCG algorithm for the initial computation of eigenvalues. Then, for $j = 1, \dots, J$ the LOBPCG method is started with the interpolation of the solution of the next coarser level, and for T_j and for the projection we use a multigrid preconditioner.

The multigrid preconditioner T_J for A_J^δ is defined recursively. In order to obtain a condition number of $T_J A_J^\delta$ independent of $\delta > 0$, we use a hybrid smoother proposed by Hiptmair [2]: we set $T_0 = (A_0^\delta)^{-1}$ (using the parallel direct solver) and define T_j, $j = 1, \dots, J$ by

$$\mathrm{id} - A_j^\delta T_j = \big(\mathrm{id} - A_j^\delta I_j T_{j-1} I_j'\big)\big(\mathrm{id} - \delta^{-1} A_j^\delta S_j D_j S_j'\big)^\nu \big(\mathrm{id} - A_j^\delta R_j\big)^\mu. \quad (16)$$

The construction of T_j depends on the prolongation operator $I_j : \mathbf{V}_{j-1} \longrightarrow \mathbf{V}_j$ and its adjoint operator I_j' (restriction), a smoother $R_j : \mathbf{V}_j' \longrightarrow \mathbf{V}_j$ for A_j^δ, a smoother $D_j : Q_j' \longrightarrow Q_j$ for C_j, a transfer operator $S_j : Q_j \longrightarrow \mathbf{V}_j$ defined by $S_j q_j = \nabla q_j$, and its adjoint operator S_j'. The application of the operator is summarized in Algorithm 4.

4 Parallel Results

In our first parallel performance test, we consider the parallel scaling behavior of the LDL^T decomposition on a regular tetrahedral mesh in a cube $\Omega = (-1, 1)^3$. For the distribution of the tetrahedra on $P = 2^S$ processors, we use a recursive coordinate bisection (RCB) method.

In Table 1 we present the computation time of the LDL^T decomposition of the resulting Laplace problem and the Maxwell problem on different refinement levels. Note that the sequential decomposition has the complexity $O(N^{7/3})$. We observe

Table 1 Factorization time [s] of the Laplace problem (left) and the Maxwell problem (right) on refinement level l on $P = 2^S$ processors

S	d.o.f.	17 969	137 313	1 073 345
	P	$l = 4$	$l = 5$	$l = 6$
2	4	4.42	227.57	
3	8	1.61	110.23	
4	16	0.98	56.09	
5	32	1.09	27.47	
6	64	1.28	16.66	1004
7	128		16.00	770
8	256		27.25	380
9	512			289

S	d.o.f.	119 344	936 032	7 413 952
	P	$l = 4$	$l = 5$	$l = 6$
3	8	56.62		
4	16	26.71		
5	32	12.60	861	
6	64	7.43	430	
7	128	9.11	309	
8	256		174	
9	512		130	5079
10	1024		261	4382

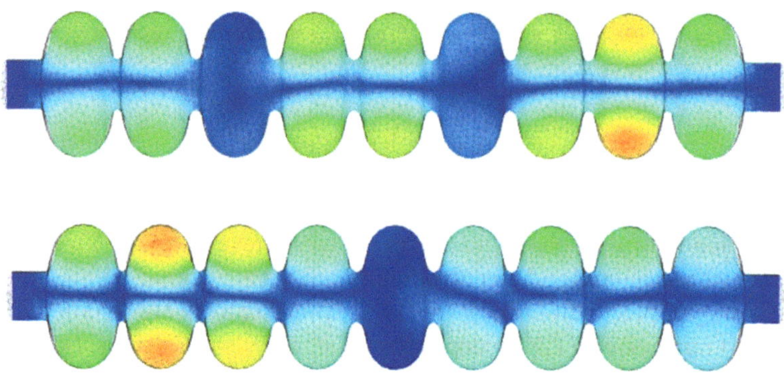

Fig. 1 Two eigenfunctions of the particle accelerator

Table 2 Degrees of freedom (dof) of Laplace- and Maxwell-problem and computed eigenmodes of the particle accelerator

Level	#dof Laplace	#dof Maxwell	Mode 1	Mode 2	Mode 3	Mode 4	Mode 5	Mode 6
0	25 816	164 324	1.2992	1.3152	1.3161	1.3256	1.3284	1.3385
1	190 140	1 266 711	1.2884	1.2950	1.2976	1.3029	1.3091	1.3123
2	1 456 851	9 943 426	1.2848	1.2892	1.2913	1.2913	1.2954	1.2976
3	11 400 277	78 788 884	1.2833	1.2844	1.2870	1.2904	1.2944	1.2983

nearly optimal scaling behavior for large problems up to 256 processors. Since the communication part in our algorithm dominates for small problems, the total computing time grows for more processors.

For an application test, we choose the geometry of a resonator in a particle accelerator. In Fig. 1 the third and eighth eigenfunction of this problem is shown.

In Table 2 we present the first six computed eigenmodes of the particle accelerator for different refinement levels. Here, we start the LOBPCG method for 30 modes with $\delta = 0.002$ and we stop the iteration if the residual for the first 10 modes is smaller than 0.001. Then, the nested eigenvalue solver requires at most 4 iterations on every level. On 64 processors the decomposition time for the coarse problems is about 30 sec and the application requires less than 2 sec. On level 2, the overall computation time is less than one hour and requires more than 100 applications of the multigrid preconditioner T_h and the same number of nearly exact Laplace solutions for the evaluation of the projection. On level 3 and 512 processor kernels the scaling is not optimal and will be a topic of future research.

Acknowledgements The authors acknowledge the financial support from BMBF grant 01IH08014A within the joint research project ASIL (Advanced Solvers Integrated Library). We thank W. Koch (CST) for the construction of the accelerator configuration.

References

1. Amestoy, P.R., Duff, I.S., Koster, J., L'Excellent, J.-Y.: A fully asynchronous multifrontal solver using distributed dynamic scheduling. SIAM J. Matrix Anal. Appl. **23**(1), 15–41 (2001)
2. Hiptmair, R.: Multigrid method for maxwell's equations. SIAM J. Numer. Anal. **36**(1), 204–436 (1998)
3. Knayzev, A.V.: Toward the optimal preconditioned eigensolver: locally optimal block preconditioned conjugate gradient method. SIAM J. Scientific Comput. **23**(2), 517–541 (2001)
4. Maurer, D., Wieners, C.: A parallel block LU decomposition method for distributed finite element matrices. Parallel Comput. http://dx.doi.org/10.1016/j.parco.2011.05.007 (2011)
5. Monk, P.: Finite Element Methods for Maxwell's Equations. Clarendon, Oxford (2003)
6. Schenk, O., Gärtner, K.: Solving unsymmetric sparse systems of linear equations with PARDISO. Sloot, P.M.A. et al. (eds.) Computational science – ICCS 2002. 2nd international conference, Amsterdam, the Netherlands, April 21–24, 2002. Proceedings. Part 2. Berlin, Springer. Lect. Notes Comput. Sci. **2330**, 355–363 (2002)
7. Schenk, O., Gärtner, K., Fichtner, W., Stricker, A.: PARDISO: A high-performance serial and parallel sparse linear solver in semiconductor device simulation. FGCS. Future Generat. Comput. Syst. **18**(1), 69–78 (2001)
8. Wieners, C.: Distributed point objects. A new concept for parallel finite elements. Kornhuber, R., et al. (eds.) Domain decomposition methods in science and engineering. Selected papers of the 15th international conference on domain decomposition, Berlin, Germany, July 21–25, 2003. Berlin, Springer. Lect. Notes Comput. Sci. Eng. **40**, 175–182 (2005)
9. Wieners, C.: A geometric data structure for parallel finite elements and the application to multigrid methods with block smoothing. Comput. Vis. Sci. **13**(4), 161–175 (2010)
10. Zaglmayr, S.: High order finite element methods for electromagnetic field computation. PhD thesis, Johannes Kepler Universität Linz (2006)

Scaling Algebraic Multigrid Solvers: On the Road to Exascale

Allison H. Baker, Robert D. Falgout, Todd Gamblin, Tzanio V. Kolev, Martin Schulz, and Ulrike Meier Yang

Abstract Algebraic Multigrid (AMG) solvers are an essential component of many large-scale scientific simulation codes. Their continued numerical scalability and efficient implementation is critical for preparing these codes for exascale. Our experiences on modern multi-core machines show that significant challenges must be addressed for AMG to perform well on such machines. We discuss our experiences and describe the techniques we have used to overcome scalability challenges for AMG on hybrid architectures in preparation for exascale.

1 Introduction

Sparse iterative linear solvers are critical for large-scale scientific simulations, many of which spend the majority of their run time in solvers. Algebraic Multigrid (AMG) is a popular solver because of its linear run-time complexity and its proven scalability in distributed-memory environments. However, changing supercomputer architectures present challenges to AMG's continued scalability.

Multi-core processors are now standard on commodity clusters and high-end supercomputers alike, and core counts are increasing rapidly. However, distributed-memory message passing implementations, such as MPI, are not expected to work efficiently with more than hundreds of thousands of tasks. With exascale machines expected to have hundreds of millions or billions of tasks and hundreds of tasks per node, programming models will necessarily be hierarchical, with local shared-memory nodes in a larger distributed-memory message-passing environment.

With exascale in mind, we have begun to focus on a hybrid programming model for BoomerAMG [11], the AMG solver in the *hypre* [12] library. BoomerAMG

A.H. Baker · R.D. Falgout · T. Gamblin · T.V. Kolev · M. Schulz · U.M. Yang (✉)
Lawrence Livermore National Laboratory, Center for Applied Scientific Computing
e-mail: abaker@llnl.gov; rfalgout@llnl.gov; tgamblin@llnl.gov; tzanio@llnl.gov; schulzm@llnl.gov; umyang@llnl.gov

C. Bischof (eds.), *Competence in High Performance Computing 2010*,
DOI 10.1007/978-3-642-24025-6_18,

has demonstrated good weak scalability in distributed-memory environments, such as on 125,000 processors of BG/L [8], but our preliminary study [4] has shown that non-uniform memory access (NUMA) latency between sockets, deep cache hierarchies, multiple memory controllers, and reduced on-node bandwidth can be detrimental to AMG's performance.

To achieve high performance on exascale machines, we will need to ensure numerical scalability and an efficient implementation as core counts increase, memory capacity per core decreases, and on-node cache architectures become more complex. Some components of AMG that lead to very good convergence do not parallelize well or depend on the number of processors. We examine the effect of high level parallelism involving large numbers of cores on one of AMG's most important components, smoothers. We also investigate an OpenMP/MPI implementation of AMG, and its performance on three supercomputers with different node architectures: a cluster with four quad-core AMD Opteron processors, a Cray XT5 machine with two hex-core AMD Opteron processors, and a BlueGene/P system with a single quad-core PowerPC processor per node. The techniques used in these environments have broad applicability beyond AMG and will enable other solvers and simulation codes to prepare for exascale.

The remainder of this paper is organized as follows. In Sect. 2, we give an overview of the AMG method and detail our strategy to run efficiently at exascale. Section 3 describes mathematical and computational challenges of exascale for AMG smoothers. We describe the necessary steps to achieve good OpenMP performance on a multi-core node in Sect. 4. Finally, we demonstrate the performance of our hybrid BoomerAMG implementation on the three multi-core architectures in Sect. 5. Section 6 presents our conclusions.

2 The Algebraic Multigrid Solver

Multigrid (MG) linear solvers are particularly well-suited to parallel computing because their computational cost is linearly dependent on the problem size. This optimal property, also referred to as algorithmic scalability, means that proportionally increasing both the problem size and the number of processors (i.e., weak scaling), results in a roughly constant number of iterations to solution. Therefore, unsurprisingly, multigrid methods are currently quite popular for large-scale scientific computing and will play a critical role in enabling simulation codes to perform well at exascale.

2.1 Overview

An MG method's low computational cost results from restricting the original linear system to increasingly coarser grids, which require fewer operations than the fine grid. An approximate solution is determined on the coarsest grid, typically with a direct solver, and is then interpolated back up to the finest grid. On each grid level an inexpensive smoother (e.g., a simple iterative method like Gauss-Seidel) is applied. The process of starting on the fine grid, restricting to the coarse grid, and interpolating back to fine grid again is called a "V-cycle", which corresponds to a single MG iteration.

MG has two phases: *setup* and *solve*. The primary computational kernels in the setup phase are the selection of the coarse grids, creation of the interpolation operators, and the representation of the fine grid matrix operator on each coarse grid. The primary computational kernels in the solve phase are a matrix-vector multiply (MatVec) and the smoothing operator, which may closely resemble a MatVec.

AMG is a flexible and unique type of MG method because it does not require geometric grid information. In AMG, coarse "grids" are simply subsets of the fine grid variables, and the coarsening and interpolation algorithms make use of the matrix entries to select variables and determine weights. These algorithms can be quite complex, particularly in parallel. More detailed information on AMG may be found in either [8] or [14].

2.2 *Scaling Strategy for AMG*

A well-designed AMG method is algorithmically scalable in that the number of iterations should stay fixed with increasing problem size. However, an effective AMG code must also be computationally scalable: the run times should stay constant with weak scaling. Therefore, both the algorithmic details related to the underlying mathematics (which impact the convergence rate) and the implementation details of the algorithm are important. To prepare our code for exascale computing, we have begun to examine the primary components of AMG, beginning with the solve phase, to determine what issues will need to be addressed.

The BoomerAMG code was originally developed with MPI in mind (as OpenMP was not competitive at the time). As a first step towards exascale, we have focused on incorporating OpenMP more fully into the code because we found performance on multi-core clusters using only MPI to be poor [4]. As discussed in Sect. 4, the initial OpenMP results were disappointing and required careful management of memory and threads to achieve good performance. Because of its simplicity and importance, the MatVec kernel was a natural focus for these initial efforts into investigating hybrid MPI + OpenMP performance. MatVec dominates the solve phase time (approximately 60%) as it is used for both restricting and interpolating the error, computing the residual, and, if AMG is used as a preconditioner, for the Conjugate Gradient (CG) or GMRES iteration step. From an implementation perspective, the matrix is stored in a compressed sparse row (CSR) parallel data structure [7]. MatVec is threaded at the loop level such that each thread operates on a subset of its process' rows.

The other key component of the solve phase is the application of the smoother, which typically constitutes about 30% of the solve time. The smoother is critical because its effectiveness in reducing the error impacts the overall MG convergence rate, directly impacting the solution time. The challenge in designing a good parallel smoother is that Gauss-Seidel, the most popular MG smoother, is inherently sequential. Therefore in parallel AMG, a so-called hybrid Gauss-Seidel, which performs Gauss-Seidel within each task and delays updates across tasks, is typically employed. The convergence of hybrid Gauss-Seidel may be impacted by exascale computing due to the use of millions of concurrent tasks, as well as memory limitations requiring smaller problem sizes per processor. In addition, the use of OpenMP could affect convergence as loop-level threading in the smoother further partitions each process' domains. We discuss these concerns in detail in Sect. 3.

The AMG setup phase time is non-negligible and problem dependent; in some cases, depending on the number of iterations required for convergence, it can rival the time of the solve phase. Our preparation of the AMG setup phase for exascale computing is a work in progress. In particular, the coarsening and interpolation algorithms may be quite complicated in parallel [5], and the long-distance variety [6] require a sizable amount of point-to-point communications. At this point, the interpolation routines in BoomerAMG are only partially threaded due to complexity, and none of the coarsening routines use any threading at all. The third setup phase component, determining coarse grid operators via a triple matrix product, is completely threaded. However these coarse grid operators become far less-sparse than the fine grid matrix. The increasing operator density increases the number of communication pairs, which significantly impacts the MatVec time on the coarsest levels [4]. Future work for the setup phase will include the development of more thread-friendly data structures for the interpolation and coarsening routines and the investigation into ways to reduce communication on coarser grid levels, including the storage of redundant data when coarsening or the use of coarse grid operators that are not the result of a triple matrix product (i.e., non-Galerkin).

3 Smoothers

The smoothing process is at the heart of the AMG algorithm, and the quality of the smoother directly affects the design and the scalability of the multigrid solver. For a linear system with a symmetric and positive definite (SPD) matrix A, a smoother is another matrix M such that the iteration

$$e^0 = e\,, \qquad e^{n+1} = (I - M^{-1}A)e^n \quad \text{for } n = 0, 1, \ldots$$

reduces the high-frequency components of an initial error vector e. This makes the remaining error smooth, so it can be handled by the coarse grid corrections. The smoother should also be convergent, so that the above iteration does not increase the low-frequency components of the error. When M is not symmetric, one can consider the symmetrized smoother $\widetilde{M} = M^T(M^T + M - A)^{-1}M$, which corresponds to a smoothing iteration with M, followed by a pass with M^T. The symmetrized smoother is often used when preconditioning CG with AMG, since CG requires a symmetric preconditioner.

One classical example of a convergent smoother is the Gauss-Seidel (GS) method, which is obtained by setting $M_{GS} = L + D$, where L is the lower triangular part and D is the diagonal part of A. Note that M_{GS} is not symmetric, so GS is frequently symmetrized in practice. Another class of general smoothers are the polynomial methods, where M is defined implicitly from $I - M^{-1}A = p(A)$ where p is a polynomial satisfying $p(0) = 1$. Both of these smoothing approaches have been essential in the development of serial algebraic multigrid and have excellent smoothing properties.

Parallel architectures, however, present serious challenges for these algorithms; GS is sequential in nature, while polynomial smoothers need knowledge of the spectrum of the matrix. Thus, a major concern for extending AMG for massively parallel machines has been the development of parallel smoothers that can maintain both scalability and good smoothing properties. In this section we summarize the theoretical and numerical results from [2] for several promising smoothers in the BoomerAMG code. Previous research in parallel smoothers can be found in [1, 15].

3.1 Hybrid Gauss-Seidel

The default smoother in BoomerAMG is a parallel version of GS known as hybrid Gauss-Seidel (hybrid-GS), which can be viewed as an inexact block-diagonal (Jacobi) smoother with GS sweeps inside each process. In other words, hybrid-GS corresponds to the block-diagonal matrix M_{HGS}, each block of which equals the process-owned $L + D$ part of A (BoomerAMG matrix storage is row-wise parallel).

Even though hybrid-GS has been successful in many applications, its scalability is not guaranteed, since it approaches Jacobi when the number of processors is large, or when the problem size per processor is small. Our strategy for addressing this issue is to investigate different variants of hybrid-GS through a qualitative smoother analysis based on the two-grid theory from [9, 10]. In particular, we proposed [2] the following criterion for smoother evaluation: if there is a constant C, independent of the problem size and the parallelization, such that $\langle \widetilde{M}x, x\rangle \le C\langle \widetilde{M}_{GS}x, x\rangle$ for any vector x, then the smoother given by M will have multigrid smoothing properties comparable to (sequential) GS. This theoretical tool is important, because it allows us to predict the scalability of various approximations of M_{GS} on current and future parallel architectures.

As one application of the theory, we showed [2] that hybrid-GS will be a good smoother when the off-processor part of the matrix rows is smaller than the diagonal in each processor. This is the case, for example, when A is diagonally dominant and each diagonal processor block has at least two non-zero entries per row. However, there are practical cases where the off-processor part of the matrix is significant, due to the problem being solved (e.g., definite Maxwell discretizations) or due to the parallel partitioning of A (e.g., due to the use of threading). In these cases, hybrid-GS will behave much worse than GS and can be divergent, even for large problem sizes per processor.

To improve the robustness of hybrid-GS, we proposed [13] and analyzed [2] the ℓ_1 Gauss-Seidel smoother (ℓ_1-GS), which corresponds to $M_{\ell_1 GS} = M_{HGS} + D^{\ell_1}$, where D^{ℓ_1} is a diagonal matrix containing the ℓ_1 norms of the off-processor part of each matrix row. This smoother has the nice property that it is convergent for any SPD matrix A. It was also shown [2] that $M_{\ell_1 GS}$ is comparable to M_{GS} with a constant $C = (1 + 4/\theta)^2$ where θ satisfies $D \geq \theta D^{\ell_1}$. Since θ depends only on the discretization stencil, for many practical problems it will be bounded away from zero, and thus $M_{\ell_1 GS}$ will result in a scalable AMG solver. To improve the performance, it is advantageous to consider parallel partitioning that lower the constant C, e.g., by using knowledge from the application in order to reduce the number and magnitude of the off-diagonal entries.

3.2 Chebyshev

As a second part of our strategy for scalable multigrid smoothers, we also explore polynomial methods where the high end of the spectrum of A is approximated with several CG iterations and a fixed scaling for the lower bound.

The theory from [9] can be applied also in this case to conclude that the best polynomial smoothers are given by shifted and scaled Chebyshev polynomials. To balance cost and performance, in practice we usually use the second order Chebyshev polynomial for $D^{-1/2} A D^{-1/2}$ (Cheby(2)). The cost of this method is comparable with the symmetrized hybrid-GS smoother.

Polynomial smoothers have the major advantage that their iterations are independent of the ordering of the unknowns or the parallel partitioning of the matrix. They also need only a matrix-vector multiply routine, which is typically finely tuned on parallel machines. These advantages, however, need to be balanced with the cost of estimating the high end of the spectrum of A. In our experience so far, this cost has not affected the scalability of Cheby(2).

3.3 Numerical Results

We illustrate the numerical performance of Cheby(2) and the symmetrized version of hybrid-GS and ℓ_1-GS with several results from [2]. The test problem describes a variable coefficient diffusion which is posed on the unit square and discretized with unstructured linear triangular finite elements, see Fig. 1. We report the iteration counts for BoomerAMG used as a two-grid solver (AMG) or a preconditioner in CG (AMG-CG) with a relative convergence tolerance of 10^{-6}.

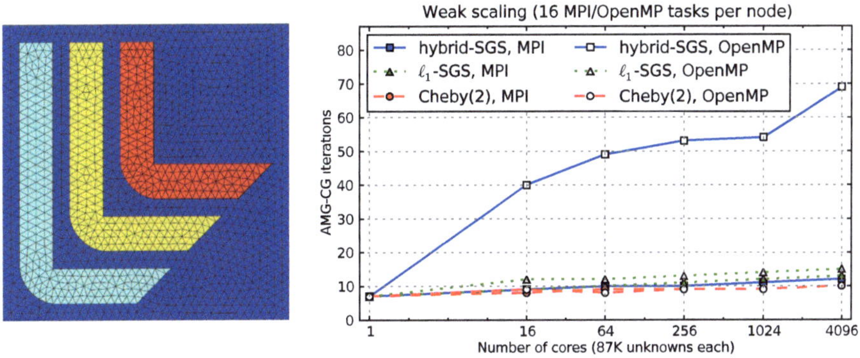

Fig. 1 Coarse mesh for the model problem with indicated material subdomains (*left*). The diffusion coefficient has 3 orders of magnitude jumps between the interior/exterior materials. Comparison of the scalability of AMG-CG with the different smoothing options when using threading (*right*)

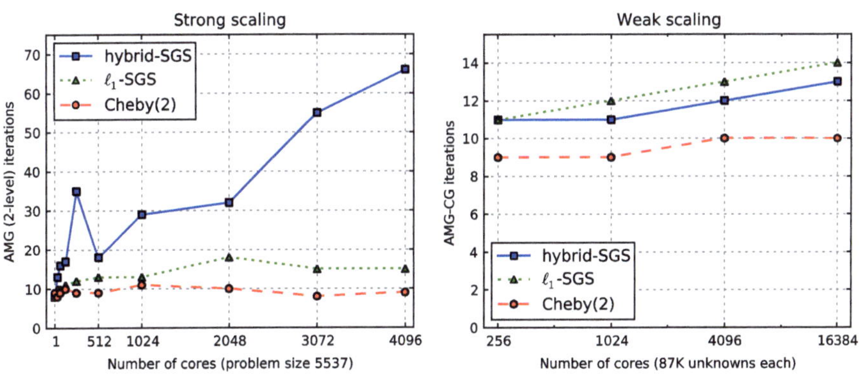

Fig. 2 Strong scaling of a two-level AMG solver with very small problem sizes per processor (*left*), and weak scalability of AMG-CG with sufficiently large problem sizes per processor (*right*)

We first investigate the impact of threading through several weak scaling runs which alternate between the use of MPI and OpenMP on compute nodes with four quad core processors each for a total of 16 cores per node. In this particular application, the numbering of the unknowns inside each processor is not guaranteed to have good locality, so the straightforward (not application-assisted) use of OpenMP introduces a bad partitioning onto the cores. As suggested by the theory, the performance of hybrid-SGS deteriorates significantly in this case, while ℓ_1-SGS and Cheby(2) remain robust. In contrast, in the MPI case the application provides parallel partitioning with a good constant θ, so all methods scale well. Note that the MPI weak scaling results on the right in Fig. 2 indicate that with application-assisted parallel partitioning, all smoothers can lead to good weak scalability on very large number of processors.

Finally, we demonstrate the impact of small problem sizes per processor through the strong scaling runs presented on the left in Fig. 2. This is an important test case, given the expected memory availability on future architectures. Since small problem sizes per processor are correlated with a large off-processor part of the matrix rows, the hybrid-SGS method deteriorates, as the smoothing analysis indicates is possible. In contrast, the effect on both ℓ_1-SGS and Cheby(2) is minimal, which is reassuring for their use on multi-core machines with small amounts of node memory.

4 On-node Threading and Memory Management

Modern High Performance Computing (HPC) systems feature increasingly complex node architectures with a rising number of compute cores per node, while the total amount of memory per node remains constant. Under such scenarios, flat programming models such as pure MPI will fail. We must provide programmers with multiple levels of concurrency, and the most common approach is a combination of MPI for cross-node communication with OpenMP for intra-node threading.

We study this setup on the Hera cluster at LLNL, a Linux cluster with 864 nodes connected by Infiniband. Each node has 16 cores distributed among four processors or sockets. Each socket features its own 8 GB memory (2 GB per core), for a total node memory of 32 GB. Any processor can access any memory location, but accesses to locations that belong to another processor's memory system incur an additional penalty. Systems such as Hera with Non-Uniform Memory Access latencies are called NUMA systems.

Figure 3 shows the speedup for two central kernels in AMG. The black line represents the MPI only version executed on a single node with varying numbers

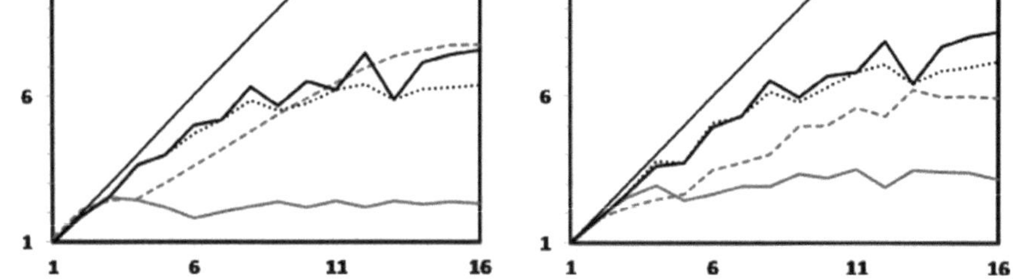

Fig. 3 Speedup for the MatVec kernel and a cycle of AMG-GMRES(10) on a single node of Hera

of cores after some minor scheduling and affinity optimizations (dotted line). Compared to this performance, the OpenMP-only version of the code (the gray solid line) performs significantly worse for all but the smallest number of cores.

On closer examination we found that the observed performance penalty is caused by the memory layout of the matrix structures in combination with the NUMA properties of the Hera nodes. Memory is allocated by the master thread and later used by all threads in the program. Since the underlying system aims to allocate all memory close to the core from which the memory is requested, all data was allocated on the memory of processor 0 leading to a large number of costly remote memory accesses from all threads running on cores in different processors as well as memory contention on processor 0.

To compensate for these problems, we developed the *MCSup* Multi-Core Support library. It provides the user with an API to allocate memory in a distributed manner across all processors in a way that matches the implicit thread distribution of OpenMP. Combined with a proactive per processor thread pinning, which is implicitly provided by *MCSup*, this technique can eliminate most remote memory accesses and reduce contention. This helps reduce the execution significantly and, as the gray dashed line in the figure shows, leads to execution times and overheads comparable to the pure MPI version.

5 Scaling of AMG

In this section we present scaling results for BoomerAMG on three different multi-core architectures: the quad-core/quad-socket Opteron cluster Hera at Lawrence Livermore National Laboratory (up to 11,664 cores), the dual hex-core Cray XT-5 Jaguar at Oak Ridge National Laboratory (up to 196,608 cores), and the quad-core Blue Gene/P system at Argonne National Laboratory (up to 1,28,000 cores). On each machine, we investigate an MPI-only version of AMG, as well as hybrid versions that use a mix of MPI and OpenMP on node. For each experiment, we utilize all available cores per node on the respective machine. We investigate the performance of AMG-GMRES(10) applied to a Laplace problem on a domain of size $N \times N \times \alpha N$, where $\alpha = 1$ on Hera and Intrepid, and $\alpha = 0.9$ on Jaguar, to allow more optimal partitioning when using 6 or 12 threads per node. The domain is decomposed in such a way that each processor has $50 \times 50 \times 25$ unknowns on Hera and Intrepid and $50 \times 50 \times 30$ on Jaguar. We consider both MPI-only and hybrid MPI/OpenMP runs and use the notation described in Figure 4. In addition, for Hera, we include a version, labeled "HmxnMC", that uses the MCSup library described in Sect. 4.

We use hybrid-GS as a smoother. The number of iterations to convergence varies across experimental setups from 17 to 44. Note that, since both the coarsening algorithm and the smoother are dependent on the number of tasks and the domain partitioning among MPI tasks and OpenMP threads, the number of iterations can vary for different combination of MPI tasks and OpenMP threads, even when using

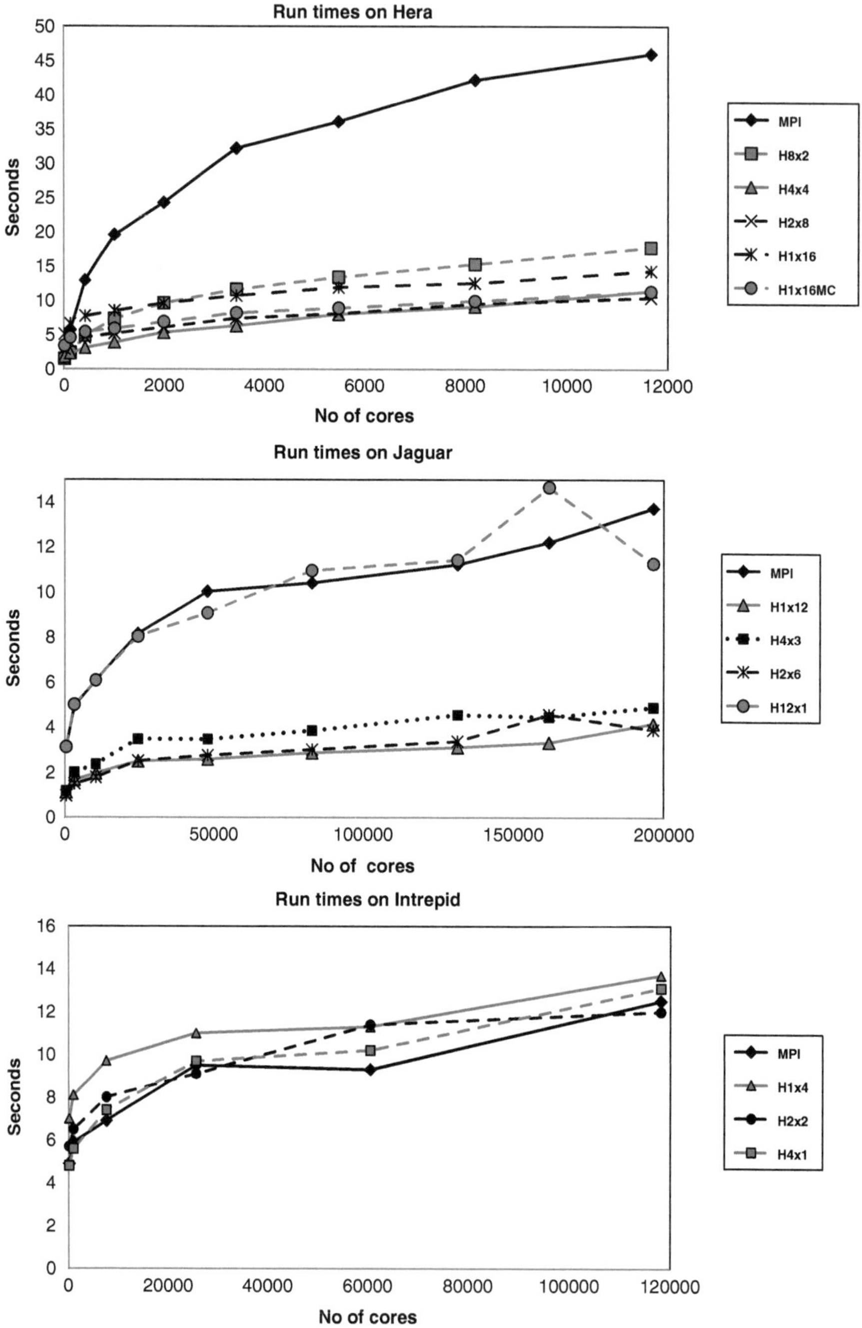

Fig. 4 Total times for AMG-GMRES(10) applied to a 7-point 3D Laplace problem on three different multi-core architectures; Hmxn denotes runs performed with m MPI tasks per cluster and n OpenMP threads per MPI task, "MPI" denotes runs performed with the MPI-only version, HmxnMC denotes the use of MCSup

the same problem size and number of cores. We present total times in Fig. 4. Separate setup and cycle times for this problem on the three architectures are described in the original paper [3].

It is immediately apparent that on the two NUMA architectures, Hera and Jaguar, the MPI-only versions as well as H12 $\times$ 1 on Jaguar, perform significantly worse than the other versions, whereas on Intrepid the MPI-only version generally performs best, with the exception of the less optimal processor geometries of 27,648 and 1,28,000 cores, where H2 $\times$ 2 is somewhat better. The worst performance on Intrepid is observed when using 4 threads per MPI task. Note that not all of the setup phase is threaded, leading to less parallelism when OpenMP is used, causing the lower performance of H1 $\times$ 4 on Intrepid, which has significantly slower cores than Jaguar or Hera. Interestingly enough this effect is not notable on Hera and Jaguar, which are however severely effected by the fact that the algorithms in the setup phase are complex and contain a large amount of non-collective communication leading to a large communication overhead and network contention. This effect is still visible, but less pronounced in the solve phase, which has a smaller amount of communication. On Hera, the worst performance for the solve phase is obtained for H1 $\times$ 16 (see [3, 4]), and is caused by the NUMA architecture. Using the *MCSup* library, see H1 $\times$ 16MC, performance is significantly improved. In the setup phase there is no NUMA effect for H1 $\times$ 16, since it mainly uses temporary storage, which is allocated within an OpenMP thread and therefore placed into the right memory module. For Hera and Jaguar initially the best times are obtained for the version that maps best to the architecture (H4 $\times$ 4 for Hera and H2 $\times$ 6 for Jaguar), to be then surpassed by H2 $\times$ 8 for Hera and H1 $\times$ 12 for Jaguar, versions with smaller network contention. Note that for the largest run on Jaguar, H1 $\times$ 12 takes more iterations than H2x6, causing H2 $\times$ 6 to be faster.

6 Conclusions

We investigated algebraic multigrid for exascale machines and considered both mathematical as well as computer science aspects to achieving scalability. Our investigation of smoothers showed that hybrid-GS promises to work well for certain problems even when we are dealing with millions or billions of cores. For more complicated problems ℓ_1-GS and polynomial smoothers are a viable fully parallel alternative to hybrid-GS because their convergence is not affected by the high level of parallelism needed for efficient implementations on exascale machines. Our tests showed that the performance of AMG varied across the three different multi-core architectures considered. A general solution is not possible without taking into account the specific target architecture. With the right settings we can achieve a performance level using a hybrid OpenMP/MPI programming model that is at least equivalent to the existing MPI model, yet has the promise to scale to core counts that prohibit the use of MPI-only applications.

Acknowledgments This work was performed under the auspices of the U.S. Department of Energy by Lawrence Livermore National Laboratory under contract DE-AC52-07NA27344. It also used resources of the Argonne Leadership Computing Facility at Argonne National Laboratory, which is supported by the Office of Science of the U.S. Department of Energy under contract DE-AC02-06CH11357, as well as resources of the National Center for Computational Sciences at Oak Ridge National Laboratory, which is supported by the Office of Science of the U.S. Department of Energy under Contract No. DE-AC05-00OR22725. These resources were made available via the Performance Evaluation and Analysis Consortium End Station, a Department of Energy INCITE project. Neither Contractor, DOE, or the U.S. Government, nor any person acting on their behalf: (a) makes any warranty or representation, express or implied, with respect to the information contained in this document; or (b) assumes any liabilities with respect to the use of, or damages resulting from the use of any information contained in the document.

References

1. Adams, M., Brezina, M., Hu, J., Tuminaro, R.: Parallel multigrid smoothing: Polynomial versus Gauss-Seidel. J. Comput. Phys. **188**, 593–610 (2003)
2. Baker, A.H., Falgout, R.D., Kolev, T.V., Yang, U.M.: Multigrid smoothers for ultra-parallel computing. 2010. (submitted). Also available as a Lawrence Livermore National Laboratory technical report LLNL-JRNL-435315
3. Baker, A.H., Gamblin, T., Schulz, M., Yang, U.M.: Challenges of scaling algebraic multigrid across modern multicore architectures. In: Proceedings of the 25th IEEE International Parallel and Distributed Processing Symposium (IPDPS 2011) (2011) To appear. Also available as LLNL Tech. Report LLNL-CONF-458074.
4. Baker, A.H., Schulz, M., Yang, U.M.: On the performance of an algebraic multigrid solver on multicore clusters. In: J.M.L.M. Palma et al, editor, *VECPAR 2010, Lecture Notes in Computer Science 6449*, pp. 102–115. Springer (2010) Berkeley, CA, June 2010. http://vecpar.fe.up.pt/2010/papers/24.php
5. Chow, E., Falgout, R., Hu, J., Tuminaro, R., Yang, U.: A survey of parallelization techniques for multigrid solvers. In: Heroux, M., Raghavan, P., Simon, H. (eds.) Parallel Processing for Scientific Computing. SIAM Series on Software, Environments, and Tools (2006)
6. De Sterck, H., Falgout, R.D., Nolting, J., Yang, U.M.: Distance-two interpolation for parallel algebraic multigrid. Num. Lin. Alg. Appl. **15**, 115–139 (2008)
7. Falgout, R., Jones, J., Yang, U.M.: Pursuing scalability for hypre's conceptual interfaces. ACM ToMS **31**, 326–350 (2005)
8. Falgout, R.D.: An introduction to algebraic multigrid. Comput. Sci. Eng. **8**(6), 24–33 (2006)
9. Falgout, R.D., Vassilevski, P.S.: On generalizing the algebraic multigrid framework. SIAM J. Numer. Anal. **42**(4), 1669–1693 (2004) UCRL-JC-150807.
10. Falgout, R.D., Vassilevski, P.S., Zikatanov, L.T.: On two-grid convergence estimates. Numer. Linear Algebra Appl. **12**(5–6), 471–494 (2005); UCRL-JRNL-203843.
11. Henson, V.E., Yang, U.M.: BoomerAMG: a parallel algebraic multigrid solver and preconditioner. Appl. Numer. Math. **41**, 155–177 (2002)
12. *hypre*. High performance preconditioners. http://www.llnl.gov/CASC/linear_solvers/
13. Kolev, T., Vassilevski, P.: Parallel auxiliary space AMG for H(curl) problems. J. Comput. Math. **27**, 604–623 (2009)
14. Stüben, K.: An introduction to algebraic multigrid. In: Trottenberg, U., Oosterlee, C., Schüller, A. (eds.) Multigrid, pp. 413–532. Academic Press, London (2001)
15. Yang, U.M.: On the use of relaxation parameters in hybrid smoothers. Numer. Linear Algebra Appl. **11**, 155–172 (2004); UCRL-JC-151575

Radiation Hydrodynamics Code LARED-H for Laser Fusion Simulation

Zeng Qinghong, Pei Wenbing, Cheng Juan, Yong Heng, and Zhai Chuanlei

Abstract LARED-H is a radiation hydrodynamics code in rz-cylindrical coordinates, developed for numerical simulation of laser inertial confinement fusion (ICF) in Institute of Applied Physics and Computational Mathematics (IAPCM). LARED-H is built on JASMIN, IAPCM's adaptive structured mesh applications infrastructure. Currently, LARED-H can accomplish the integrated simulation of ignition target. Because structured grids can not handle the complicated geometry and multi-material configuration of ICF, multi-block structured grids are employed in LARED-H. Using multi-block grids, we can deal with complicated geometry and generate initial meshes with good quality. Large deformation of fluid is one of the most difficult issues of numerical simulation of laser fusion. In LARED-H code, the strategy of "Lagrange plus remapping" is used to resolve the extreme distortion of computational meshes. We allow the meshes move with fluid until they get tangled, and then transform the physical variables from the tangled meshes to new meshes. On the new meshes, the material interface is not necessary to maintain as Lagrangian curve and is allowed to cross the cells. Therefore, mixed cells are introduced. To model the mixed cells, interface tracing algorithms of material interface and mixture models are developed. To discretize the three-temperatures energy equations, Kershaw diffusion scheme is used. In our code, Kershaw diffusion scheme is extended from structured grids to multi-block girds according to continuous flux conditions. An ignition target is simulated by LARED-H code and numerical results are demonstrated.

Z. Qinghong (✉) · P. Wenbing · C. Juan · Y. Heng · Z. Chuanlei
Institute of Applied Physics and Computational Mathematics, Beijing, P. R. China
e-mail: zeng_qinghong@iapcm.ac.cn; pei_wenbing@iapcm.ac.cn; cheng_juan@iapcm.ac.cn; yong_heng@iapcm.ac.cn; zhai_chuanlei@iapcm.ac.cn

C. Bischof (eds.), *Competence in High Performance Computing 2010*, DOI 10.1007/978-3-642-24025-6_19,

1 Introduction

Energy crisis is one of the most important problems today in the world. Inertial confinement fusion (ICF) will be part of the solution of energy problem.

ICF is a process where nuclear fusion reactions are initiated by heating and compressing a fuel target, typically in the form of a capsule that most often contains a mixture of deuterium and tritium. To compress and heat the fuel, energy is delivered to the outer layer of the target using high-energy beams of laser light, electrons or ions. The heated outer layer explodes outward, producing a reaction force against the remainder of the target, accelerating it inwards, and sending shock waves into the center. A sufficiently powerful set of shock waves can compress and heat the fuel at the center so much that fusion reactions occur.

Indirect drive laser fusion is one the major branches of ICF research. Indirect drive laser fusion uses a "hohlraum" for which the capsule is located in the center, as shown in Fig. 1. High energy laser beams are injected from the either side and are transformed into X-rays. X-rays ablate and compress the capsule for fusion. The physical phenomena in laser fusion are very complicated, numerical simulation is one of the major ways for laser fusion research and essential for the designs of such hohlraums and capsules. Due to the multi-material, large distortion, strong nonlinearity and strong discontinuousness, it is very difficult to simulate the integrated process of laser fusion.

LARED-H is a radiation hydrodynamics code in rz-cylindrical coordinates [15], developed for numerical simulation of laser fusion in Institute of Applied Physics and Computational Mathematics (IAPCM). LARED-H is built on JASMIN [8], IAPCM's adaptive structured mesh applications infrastructure, to meet the requirement for large-scale simulation of laser fusion research. Up to now, LARED-H benefiting from JASMIN has been developed largely. The underlying supports for LARED-H, including data structure, memory management, data management, and parallel computing, all come from JASMIN. Because structured grid can't handle the complicated geometry and multi-material configuration of laser fusion, multi-block structured grid is employed in LARED-H. Using multi-block grid, we

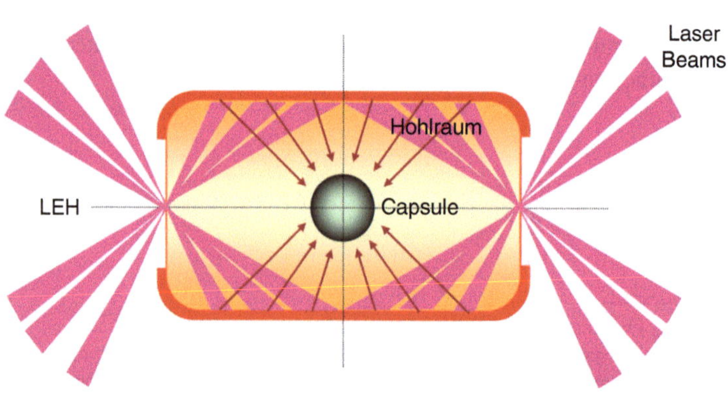

Fig. 1 Indirect drive laser fusion

can deal with complicated geometry and generate initial meshes with good quality. Currently, LARED-H can accomplish the integrated simulation of laser fusion.

2 Multi-Block Structured Grids

Structured grids are used broadly because of simple topology and data structure. But structured grids are restricted to regular computational domain. For complicated geometry domain and complicated boundaries, structured grids are difficult to generate meshes with good quality, even incompetent.

To overcome these difficulties, multi-block structured grids are employed in LARED-H code to discretize the computational domain. The computational domain is divided into a few contiguous subgeometries named by blocks. A multi-block structured grid is the union of these blocks. Each block of multi-block structured grids is structured and patched but the relation between blocks can be unstructured, see Fig. 2.

Multi-block structured grids can handle complicated geometry, complex multi-material configure and are more flexible than structured grids. According to the unstructured assembling manner, it can produce unstructured points at the corner of block. To transform the computation from structured grids to multi-block structured grids, one of major tasks is to design corresponding algorithms for these unstructured points specially, including strong connection points and degenerative connection points, as shown in Fig. 2.

3 Radiation Hydrodynamics

In laser fusion, the physical models are described by radiation hydrodynamics equations. The numerical algorithms of hydrodynamics equations and radiation energy equations used in LARED-H code will be introduced in this section.

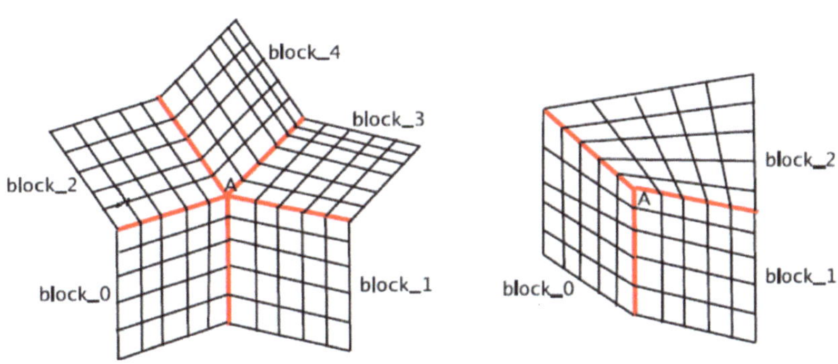

Fig. 2 Multi-block structured grids. In the left figure, there is a strong connection point which is surrounded by 5 or more cells. In the right figure, there is a degenerative connection point which is surrounded only by 3 cells.

3.1 Hydrodynamics Equations

The mass equation is

$$\frac{\mathrm{d}\rho}{\mathrm{d}t} + \rho\nabla \cdot \mathbf{u} = 0 \tag{1}$$

and momentum equations can be written as

$$\rho\frac{\mathrm{d}\mathbf{u}}{\mathrm{d}t} = -\nabla p \tag{2}$$

Here ρ is density, t is time and $\mathbf{u}$ is velocity vector.

Lagrange fluid scheme is crucial for simulation of laser fusion. Improper fluid scheme maybe produce non-physical mesh movement and one can't distinguish the non-physical movement from actually physical movement. Currently, integrated gradients algorithm [2], compatible algorithm [3, 4] and cell-centered scheme [12, 13] are all implemented in LARED-H, and in general compatible algorithm or cell-centered scheme are recommended.

Large deformation of fluid is one of the most difficult issues of numerical simulation of laser fusion. The large distortion of computational meshes can decrease the computational accuracy significantly, even break the computation down. The methods to deal with such large distortion are very important for laser fusion simulation under Lagrange or ALE frameworks. In LARED-H, the strategy of "Lagrange plus remapping" approach is used to resolve the extreme distortion of computational meshes. We allow the meshes move with fluid until they get tangled, and then transform the physical variables from the tangled meshes to new meshes. On the new meshes, the material interface is not necessary to maintain as Lagrangian curve and is allowed to cross the cells. Therefore, mixed cells (including two or more kinds of materials) are introduced. To model the mixed cells, interface reconstruction algorithms [1, 5] of material interface and mixture models [16] are developed.

3.2 Radiation Energy Equations

The radiation energy equations are modeled as three-temperatures conduction equations,

$$\rho\frac{\mathrm{d}\mathbf{e}}{\mathrm{d}t} = -\mathbf{p}\nabla \cdot \mathbf{u} + \nabla \cdot (\mathbf{k}\nabla\mathbf{T}) + \rho\mathbf{W} \tag{3}$$

where $\mathbf{e}$ is specific internal energy, $\mathbf{k}$ is the coefficient of heat conduction and $\mathbf{W}$ is source. These items can be expressed as

$$\mathbf{e} = \begin{pmatrix} e_e + e_I \\ e_i \\ e_r \end{pmatrix}, \quad \mathbf{p} = \begin{pmatrix} p_e \\ p_i + q \\ p_r \end{pmatrix}, \tag{4}$$

$$\mathbf{k}\nabla\mathbf{T} = \begin{pmatrix} k_e \nabla T_e \\ k_i \nabla T_i \\ k_r \nabla T_r \end{pmatrix}, \quad \mathbf{W} = \begin{pmatrix} w_{ei}(T_i - T_e) + w_{er}(T_r - T_e) + w_l \\ w_{ei}(T_e - T_i) \\ w_{er}(T_e - T_r) \end{pmatrix}, \tag{5}$$

Here e_e, e_i, e_r are specific internal energy of electrons, ions and radiation respectively. e_I is structure energy of electron. p_e, p_i, p_r are pressure of electrons, ions and radiation respectively. q is artificial viscosity. k_e, k_i, k_r and T_e, T_i, T_r are coefficients of heat conduction and temperature of electrons, ions and radiation respectively. w_{ei} is energy exchange rate between electrons and ions, w_{er} is energy exchange rate between electrons and radiation. w_l is energy deposit rate of laser which is obtained by using a ray tracing algorithm.

The discretization on arbitrary distortion meshes of diffusion items in radiation energy equations, i.e. $\mathbf{k}\nabla\mathbf{T}$, is very important for the numerical stability, convergence and accuracy. Kershaw diffusion scheme [6, 9, 10, 14] is employed in LARED-H for numerical discretization of these diffusion items. For arbitrary quadrilateral meshes, Kershaw scheme is a classical diffusion scheme and possesses some good properties on severely distorted meshes, but it is only suitable for quadrilateral structured meshes. In LARED-H, Kershaw scheme is extended to multi-block grids and the flux continuous condition is hold which can assure the conservation of energy [17].

After the discretization of the radiation energy equations, Picard linearization method is used to obtain the system of linear equations, and HYPRE [7] is employed to solve this system of linear equations. HYPRE is a software library of high performance preconditioners and solvers for the solution of large, sparse linear systems of equations on massively parallel computers, developed by Lawrence Livermore National Laboratory (LLNL).

4 Numerical Simulation of Ignition Target

A spherical capsule, with deuterium-tritium (D-T) gas, frozen D-T solid fuel, and an ablator, is located in a cylindrical gold hohlraum with two laser entrance holes (LEH), as shown in Fig. 3. The hohlraum peak radiation temperature (TR) is 250–300 eV, with a shaped prepulse as needed for a low-entropy implosion. The length of hohlraum is 9.5 mm, the diameter is 5.5 mm and LEH is 2.75 mm. The thickness of hohlraum is 40um and the laser pulse is described in Fig. 5.

The ignition target [11] shown in Fig. 3 is simulated by LARED-H code. Due to the symmetry, only one quarter is considered and the initial multi-block grids are shown in the top left of Fig. 4. There is a unstructured point on the symmetry axis

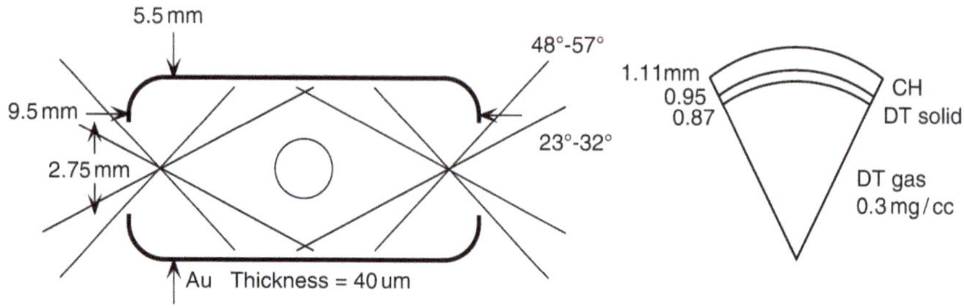

Fig. 3 An ignition target that uses 1.35 MJ of laser energy. The D-T fuel is in a cryogenic layer, surrounded by a CH ablator. The spherical capsule, shown in more detail in the right part of the figure, is suspended in the center of a cylindrical gold hohlraum, which the incoming laser beams heat to 300 eV. The beams are arranged in two cones coming in from each side

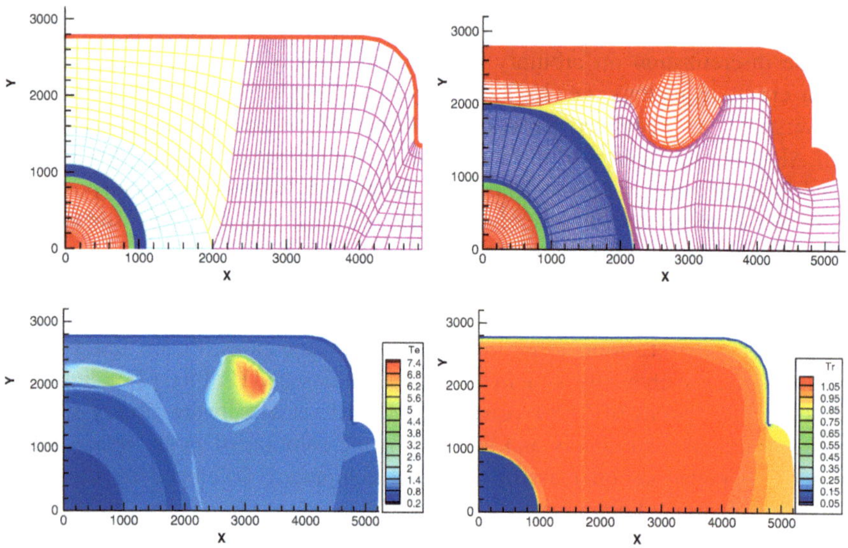

Fig. 4 Numerical results for ignition target

boundary. This unstructured point is positioned in the hohlraum and far from the capsule to avoid the influence on capsule. The Lagrange computation is continued until the computational meshes become tangled at 7.2 ns. The tangled cell which has caused the computation breaking down is not in the hohlraum or capsule, but near the LEH. The sliding between heavy gold wall and light gas filled in the hohlraum was not considered and non-physical distortion occurred near the LEH. To resolve this problem, sliding line algorithm must be developed. The meshes, electron temperature and radiation temperature at 7.2 ns are also shown in Fig. 4.

When the meshes become tangled, we then perform the remapping operation. A new meshes with good quality are reproduced and the physical variables are

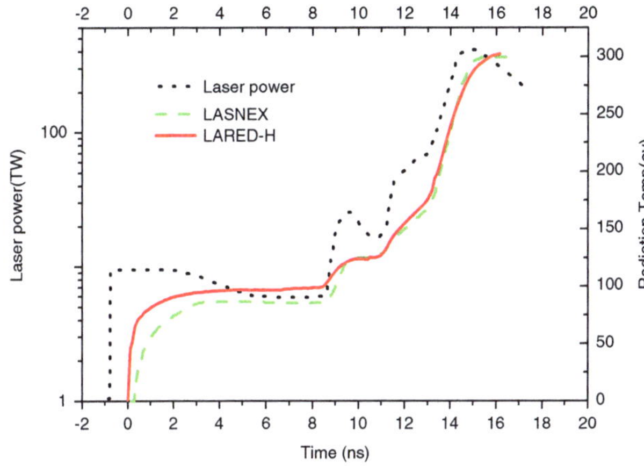

Fig. 5 Comparison of radiation temperature curves obtained from LARED-H and LASNEX. Laser power versus time to drive the ignition target (*dotted curve, left scale*), and temperature versus time optimal for the ignition capsule (*solid curve* from LARED-H and *dashed curve* from LASNEX, *right scale*). The shaped pulse prior to peak drive is needed to compress the target, increasing the pressure in a controlled way before applying peak power

transformed from the tangled meshes to new meshes, and Lagrangian computation goes on. This "Lagrange plus remapping" manner is performed repeatedly until the whole computation is finished at about 17 ns.

To validate the correctness and accuracy of LARED-H code, we compared the radiation temperature of LARED-H with that of LASNEX, shown in Fig. 5. The red solid curve is radiation temperature obtained from LARED-H, and green dashed curve is obtained from LASNEX. It can be seen that the results are comparable though a little difference is presented.

5 Conclusions and Perspectives

Up to now, LARED-H code can simulate the integrated process of the laser fusion with "Lagrange plus remapping" and mixing cell modeling. To improve the physical modeling, we are now developing LARED integration codes based on LARED-H and LARED-R for high degree of confidence. LARED-R is a radiation transport code using multi-group diffusion modeling or discrete ordinate multi-group transfer modeling.

The future works of LARED-H include the treatment of distortion of meshes in an ALE framework, multi-material interface construction, closure model and sliding line algorithm.

Acknowledgements I would like to acknowledge LARED-H team in IAPCM, this fruitful work is the result of their efforts. This work was supported by the National Basic Research Program of China under Grant No. 2005CB321702 and the National Natural Science Foundation of China under Grant No. 11001026 and 10901021.

I would also acknowledge professor Gabriel Wittum gratefully for his invitation to me to attend the CiHPC2010 conference.

References

1. Benson D.J.: Volume of fluid interface reconstruction methods for multi-material problems. Appl. Mech. Rev. **55**, 151–165 (2002)
2. Browne, P.L.: Integrated gradients: a derivation of some difference forms for the equation of motion for compressible flow in two-dimensional Lagrangian hydrodynamics, using integration of pressures over surfaces, Technical Report LA-10587-MS, Los Alamos National Laboratory (1986)
3. Caramana, E.J., Burton D.E., Shashkov M.J., Whalen P.P.: The construction of compatible hydrodynamics algorithms utilizing conservation of total energy. J. Comput. Phys. **146**, 227–262 (1998)
4. Caramana, E.J., Rousculp C.L., Burton D.E.: A compatible, energy and symmetry preserving Lagrangian hydrodynamics algorithm in three-dimensional Cartesian geometry. J. Comput. Phys. **157**, 89–119 (2000)
5. Dyadechko, V., Shashkov M.: Reconstruction of multi-material interfaces from moment data. J. Comput. Phys. **227**, 5361–5384 (2008)
6. Fatenejad, M., Moses G.A.: Extension of Kershaw diffusion scheme to hexahedral meshes. J. Comput. Phys. **227**, 2187–2194 (2008)
7. HYPRE: high performance preconditioners library. http://acts.nersc.gov/hypre/main.html.
8. JASMIN: J Adaptive Structured Mesh applications Infrastructure. http://www.iapcm.ac.cn/jasmin
9. Kershaw, D.S.: Differencing of diffusion equation in Lagrangian hydrodynamic codes. J. Comput. Phys. **39**, 375–395 (1981)
10. Krauser, W.J., Hoffman, N.M., et al.: Ignition target design and robustness studies for the National Ignition Facility. Phys. Plasmas. **3**, 2084–2093 (1996)
11. Lindl, J.D., Amendt, P., Berger, R.L, et al.: The physics basis for ignition using indirect-drive targets on the National Ignition Facility. Phys. Plasmas. **11**, 339–491 (2004)
12. Maire, P.H., Breil, J.: A second-order cell-centered Lagrangian scheme for two-dimensional compressible flow problems. Int. J. Numer. Meth. Fluids. **56**, 1417–1423 (2008)
13. Maire, P.H., Breil, J., Galera, S.: A cell-centred arbitrary Lagrangian CEulerian (ALE) method. Int. J. Numer. Meth. Fluids. **56**, 1161–1166 (2008)
14. Moses, G.A., Yuan, J.: Radiation diffusion in DRACO using Kershaw Difference scheme. Technical Report UWFDM-1213, Fusion Technology Institute, University of Wisconsin (2003)
15. Pei, W.B.: The Construction of Simulation Algorithms for Laser Fusion. Commun. Comput. Phys. **2**, 255–270 (2007)
16. Shashkov, M.: Closure models for multidimensional cells in arbitrary Lagrangian-Eulerian hydrocodes. Int. J. Numer. Meth. Fluids. **56**, 1497–1504 (2008)
17. Zeng, Q.H., Pei W.B., Cheng, J.: Extension of Kershaw diffusion scheme to multi-block grid. Defense Science and Technical Report GF-A0115352G, Institute of Applied Physics and Computational Mathematics (2009)

MIX
Papier aus verantwortungsvollen Quellen
Paper from responsible sources
FSC® C105338

If you have any concerns about our products,
you can contact us on
ProductSafety@springernature.com

In case Publisher is established outside the EU,
the EU authorized representative is:
Springer Nature Customer Service Center GmbH
Europaplatz 3, 69115 Heidelberg, Germany

Printed by Libri Plureos GmbH
in Hamburg, Germany